# This Notebook Belongs To

Name:..........................................................................

Contact:.......................................................................

Thanks for buying a copy of this book.

**A special request**

Your brief  Amazon review could really help us!

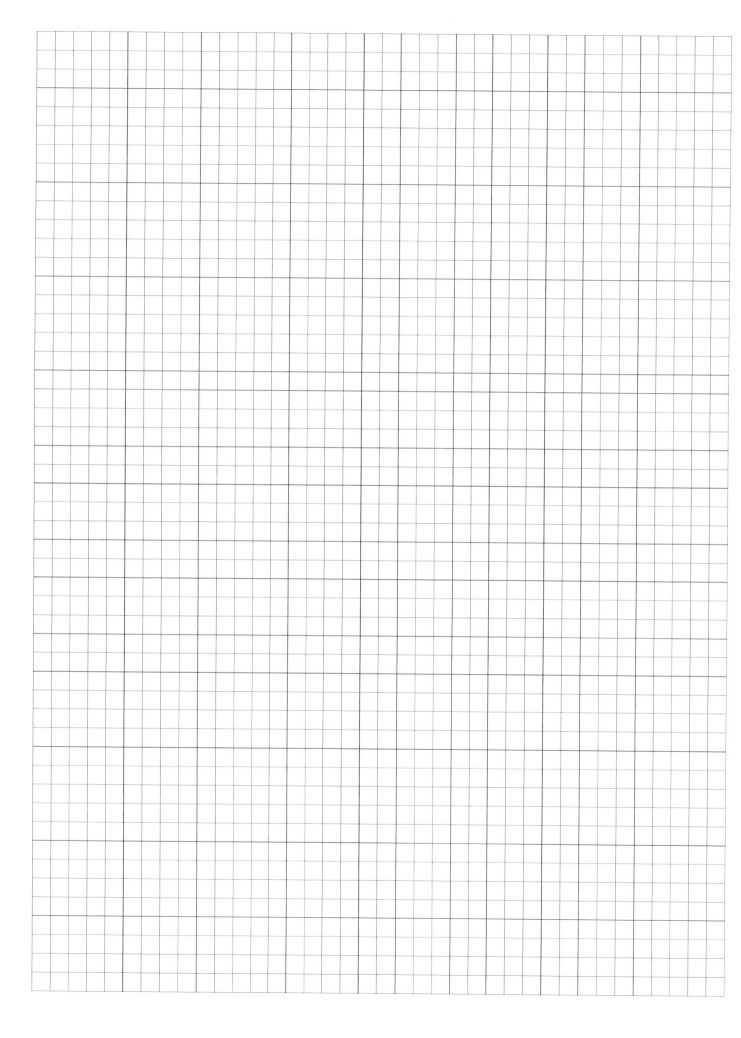

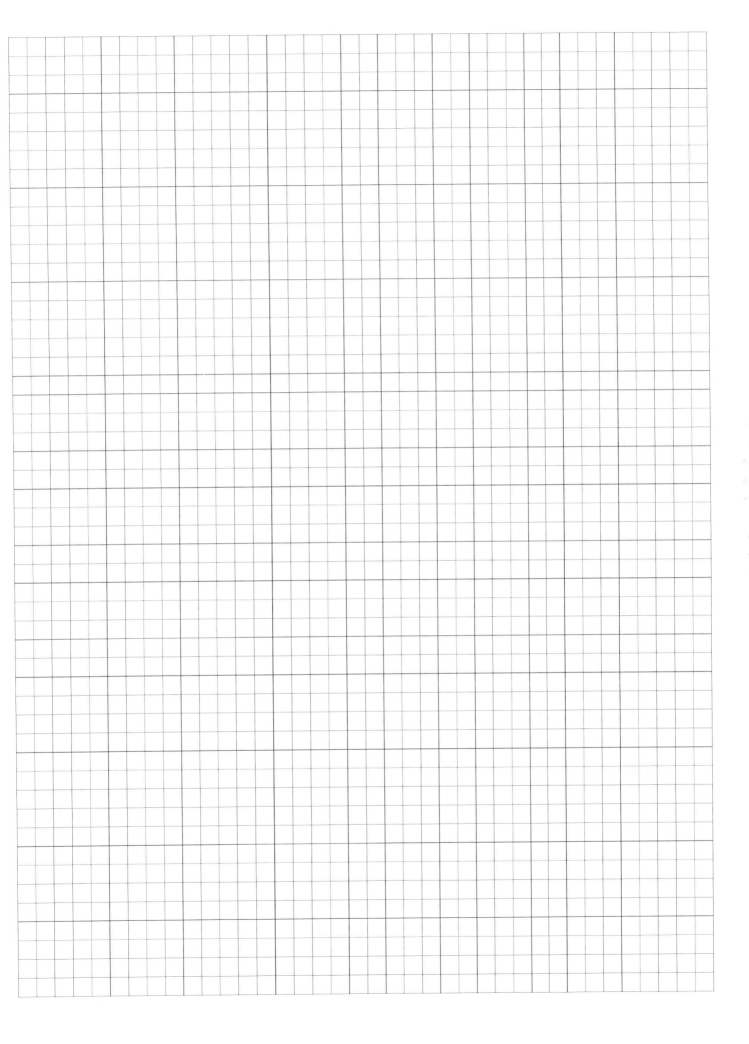

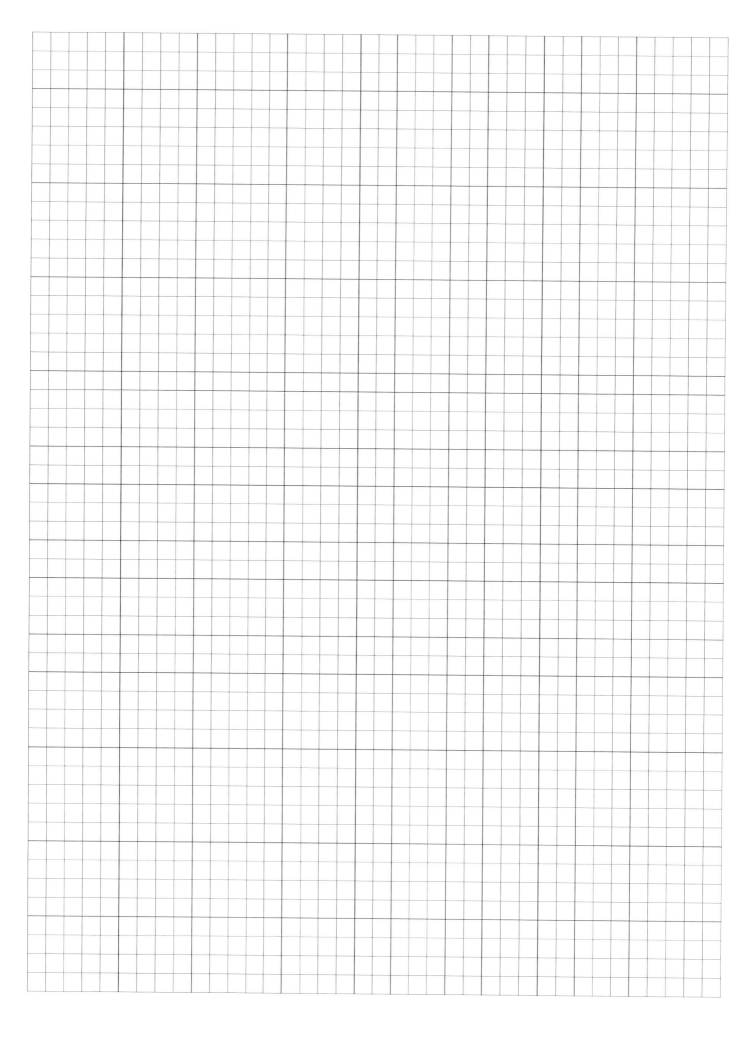

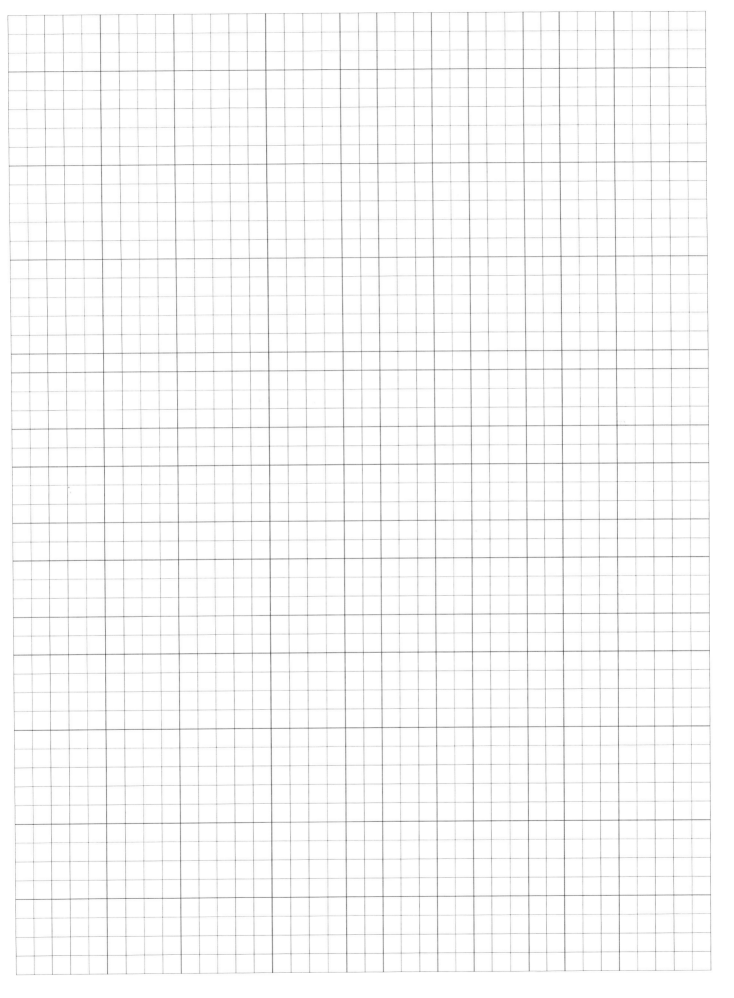

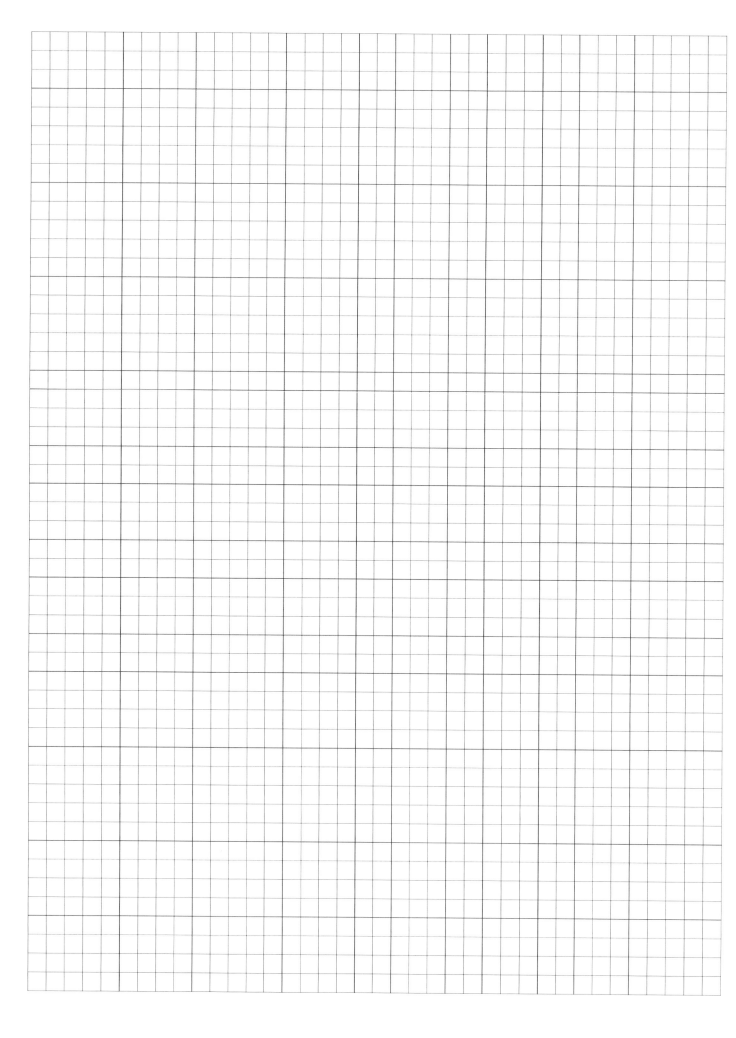

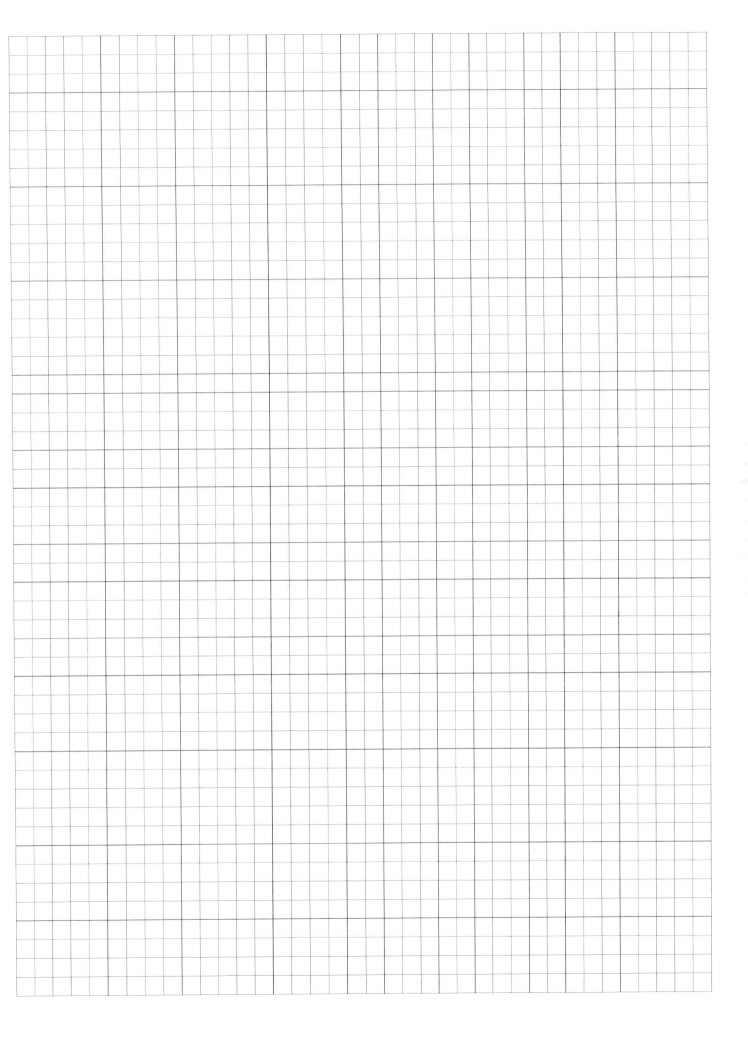

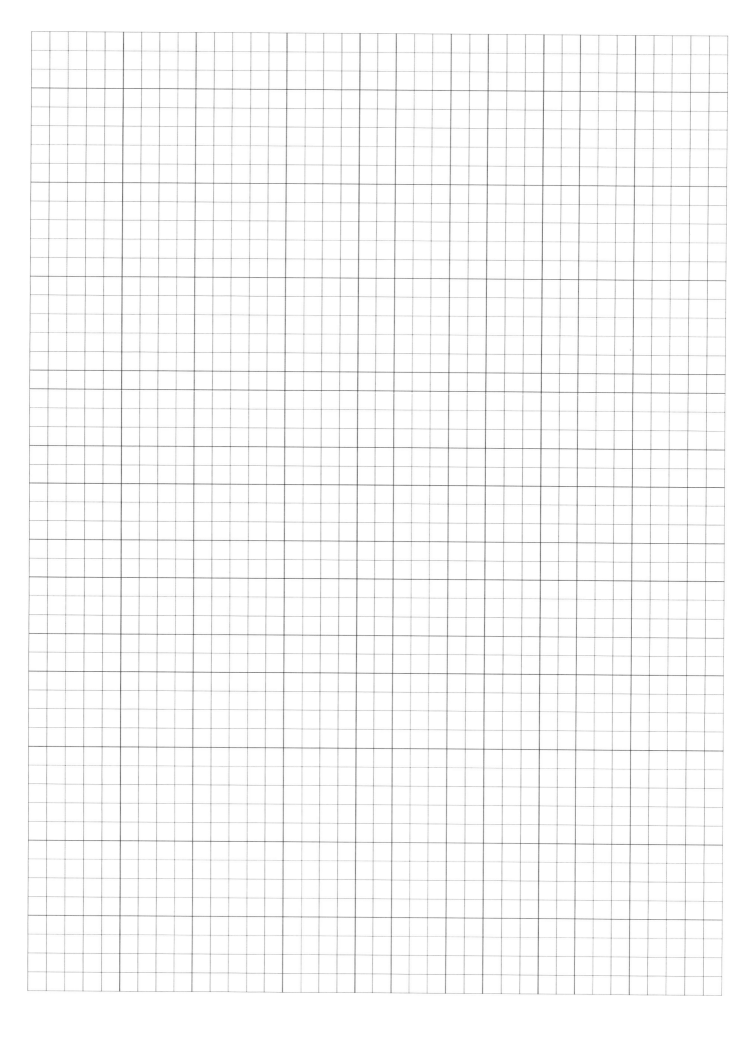

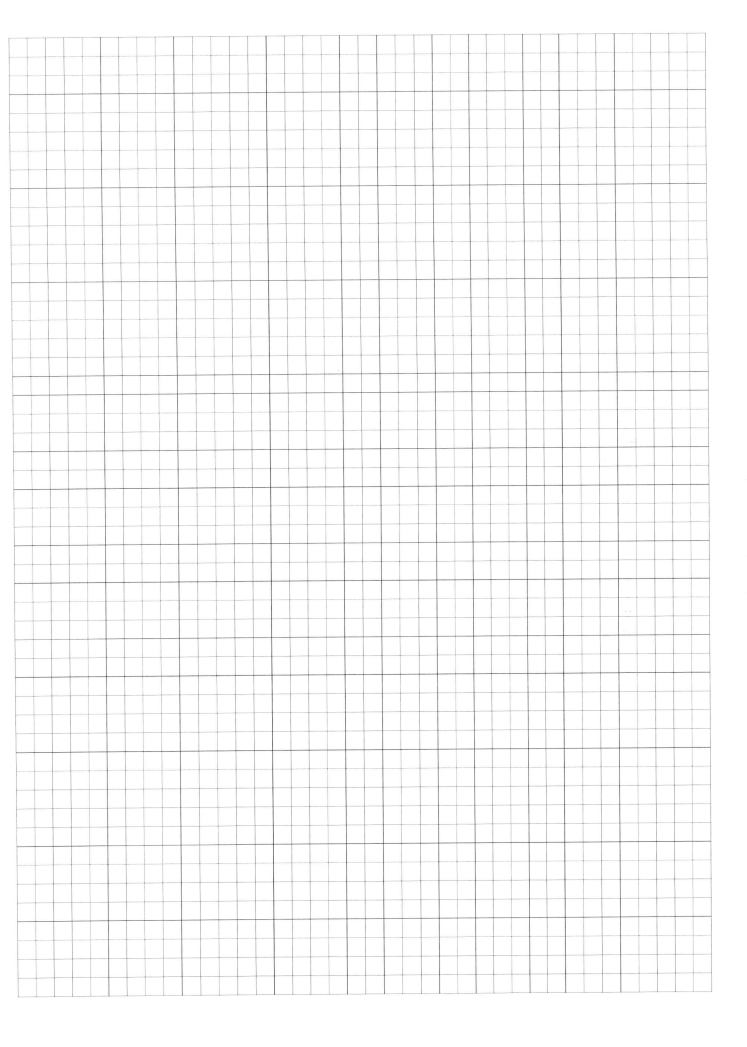

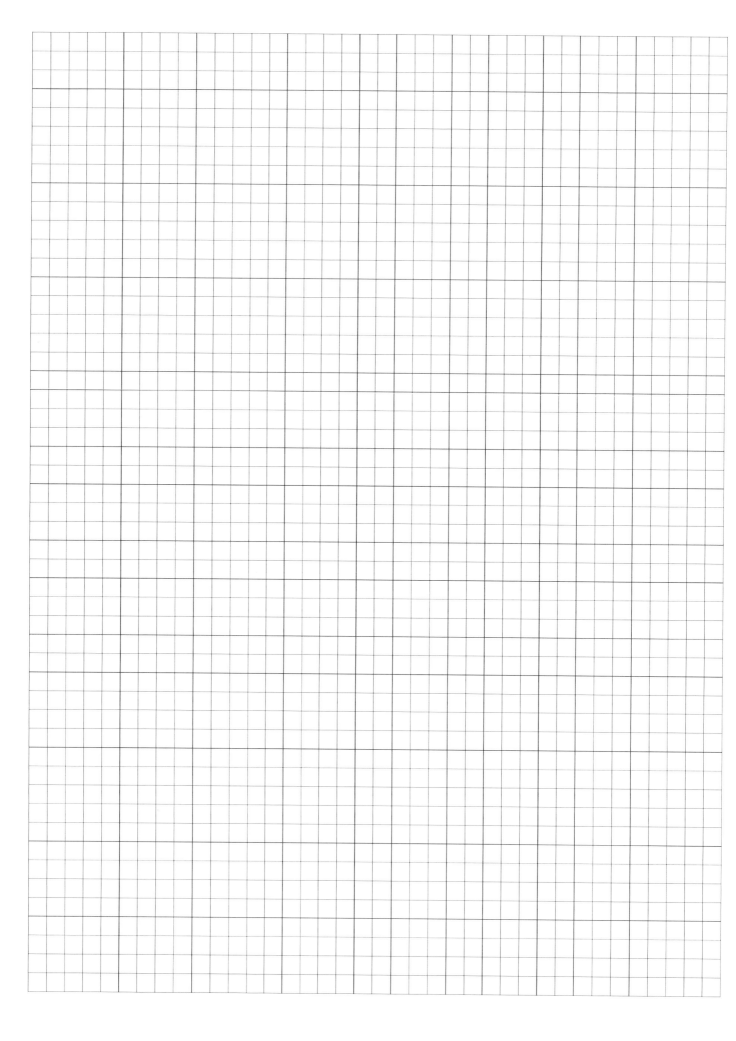

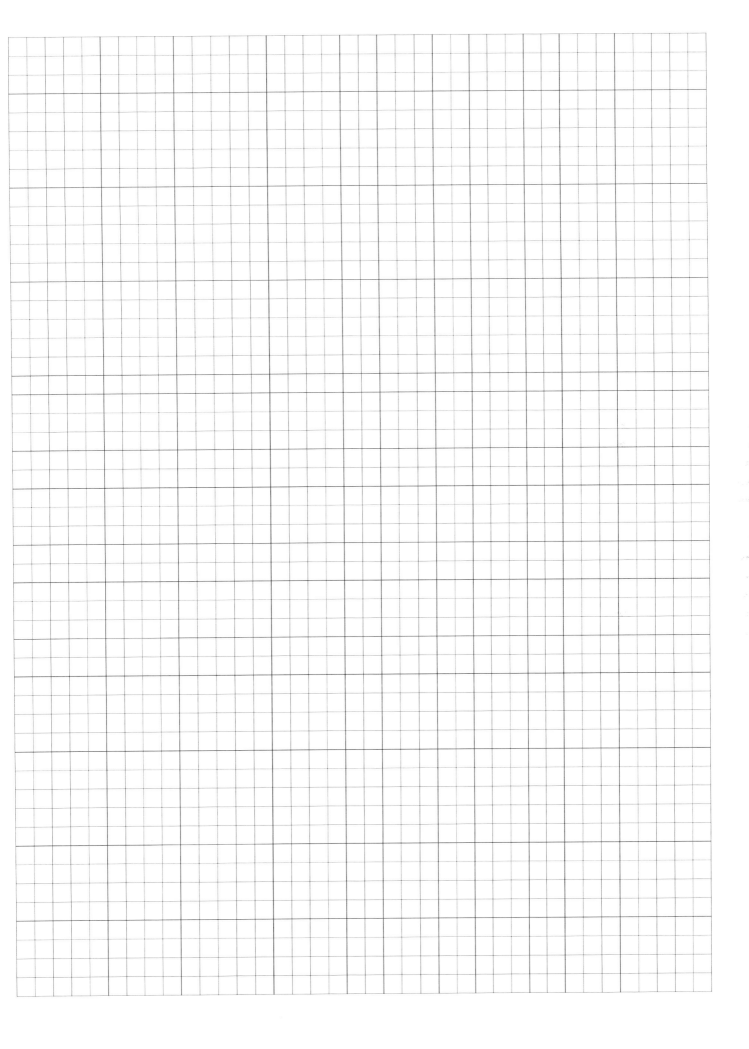

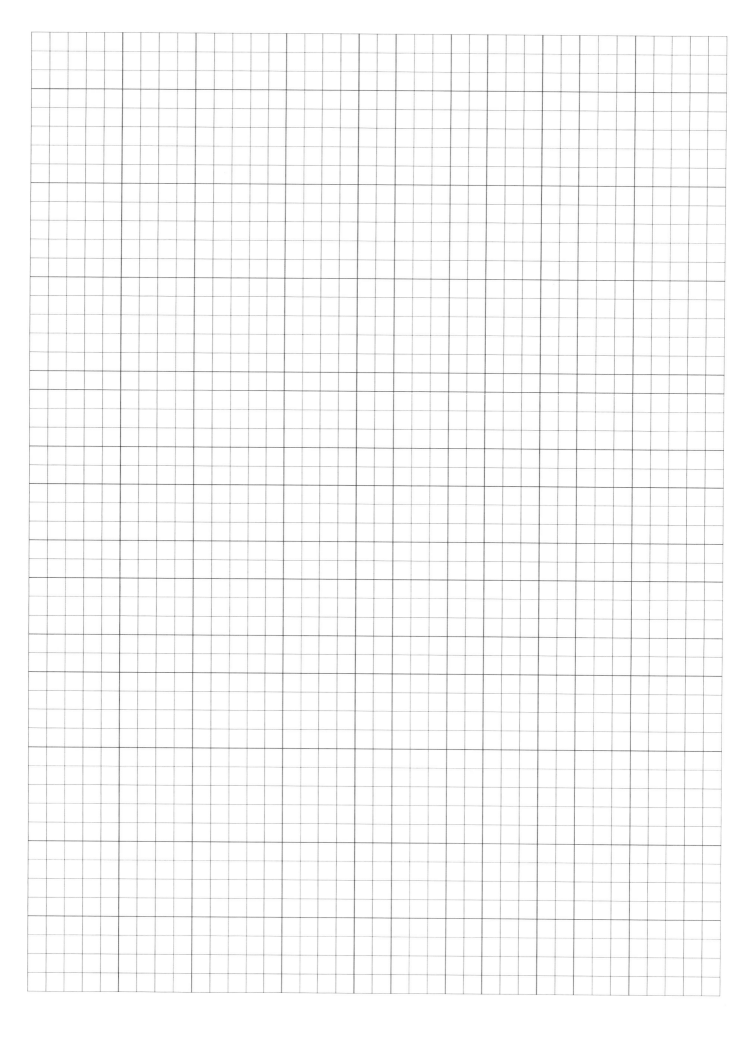

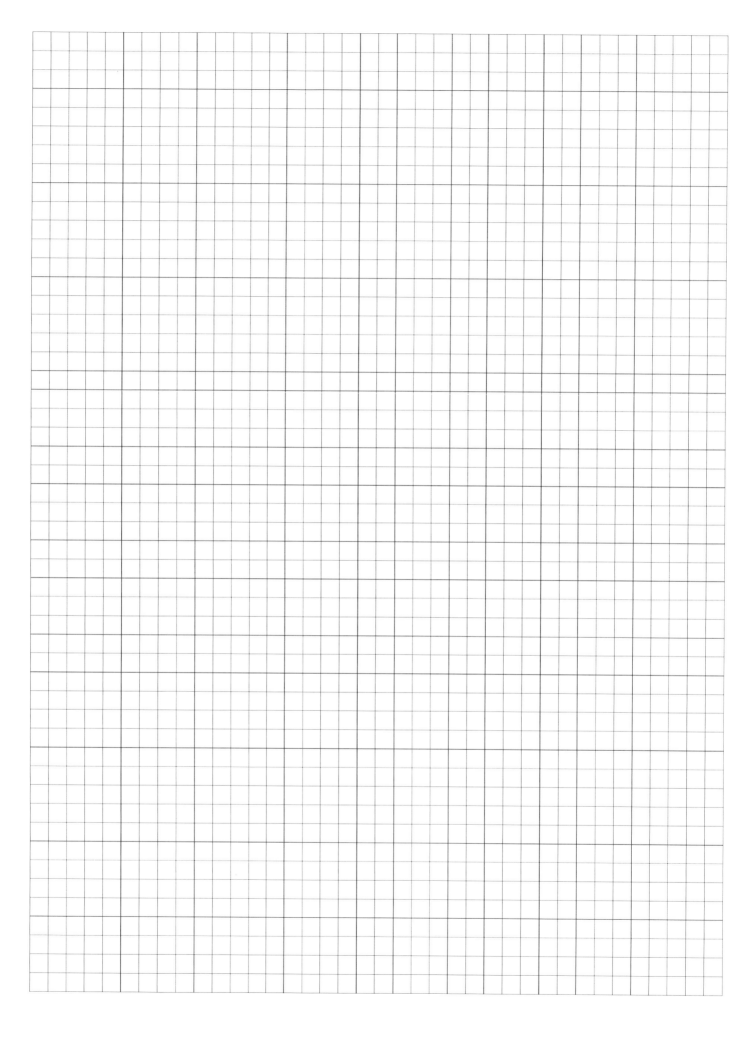

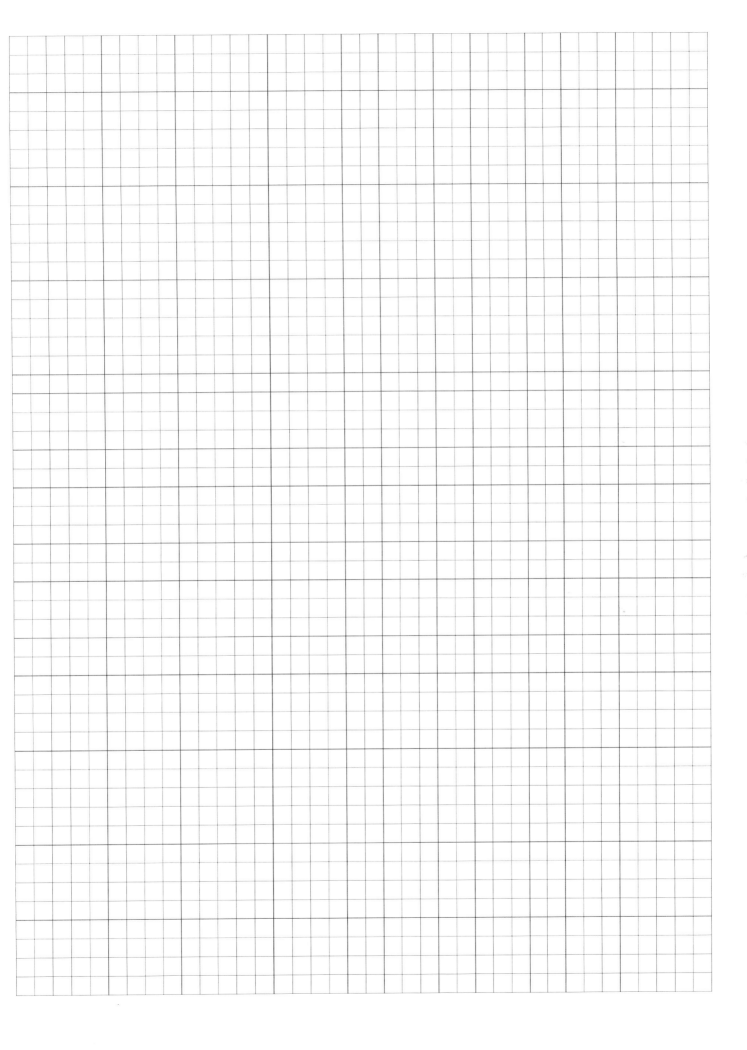

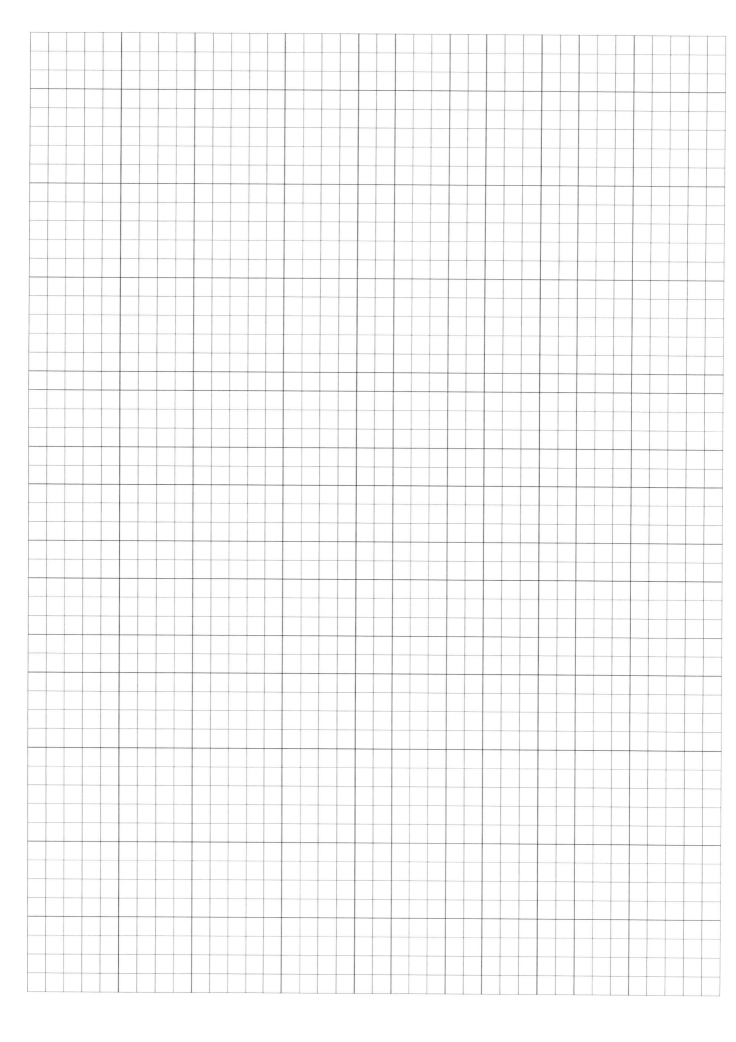

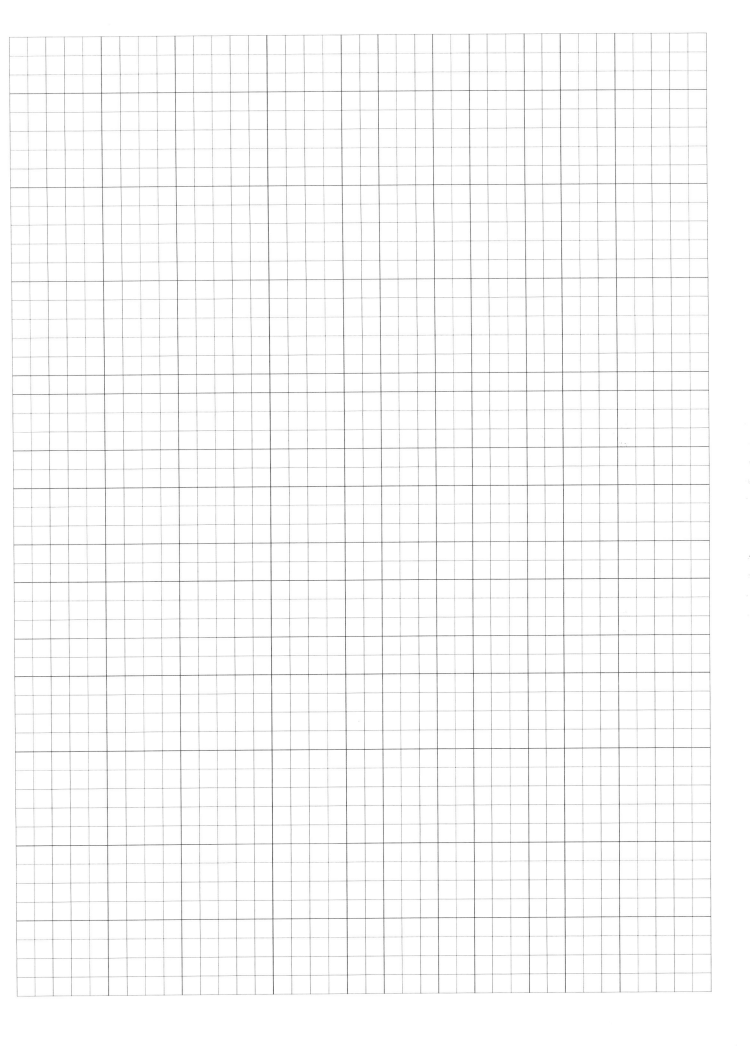

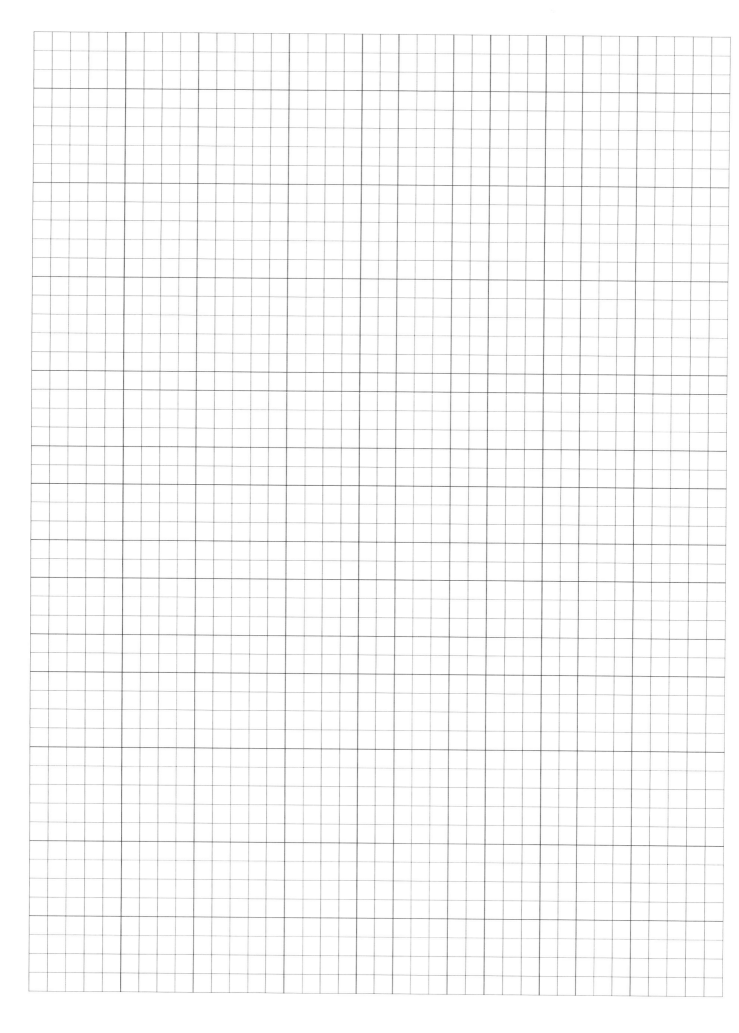

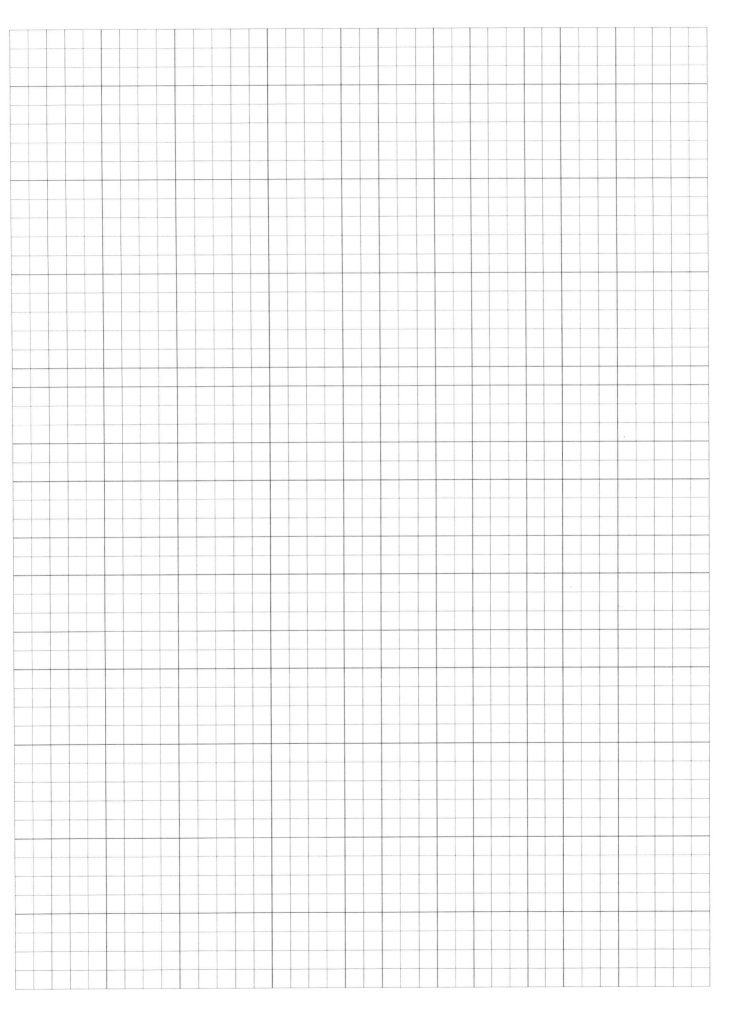

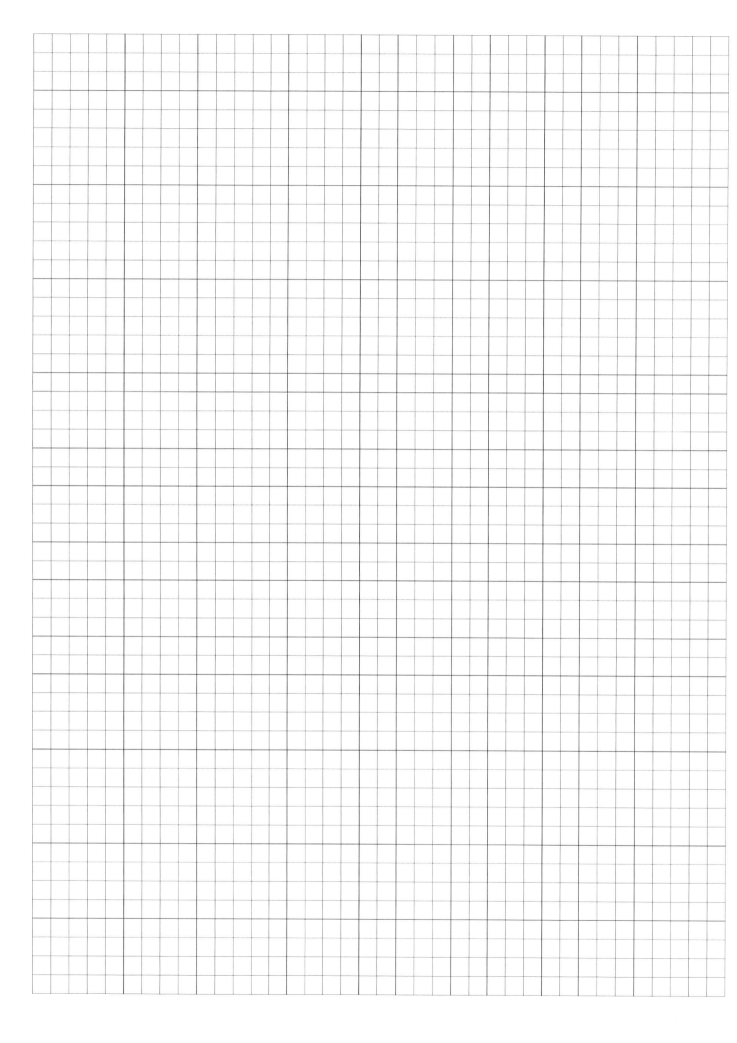

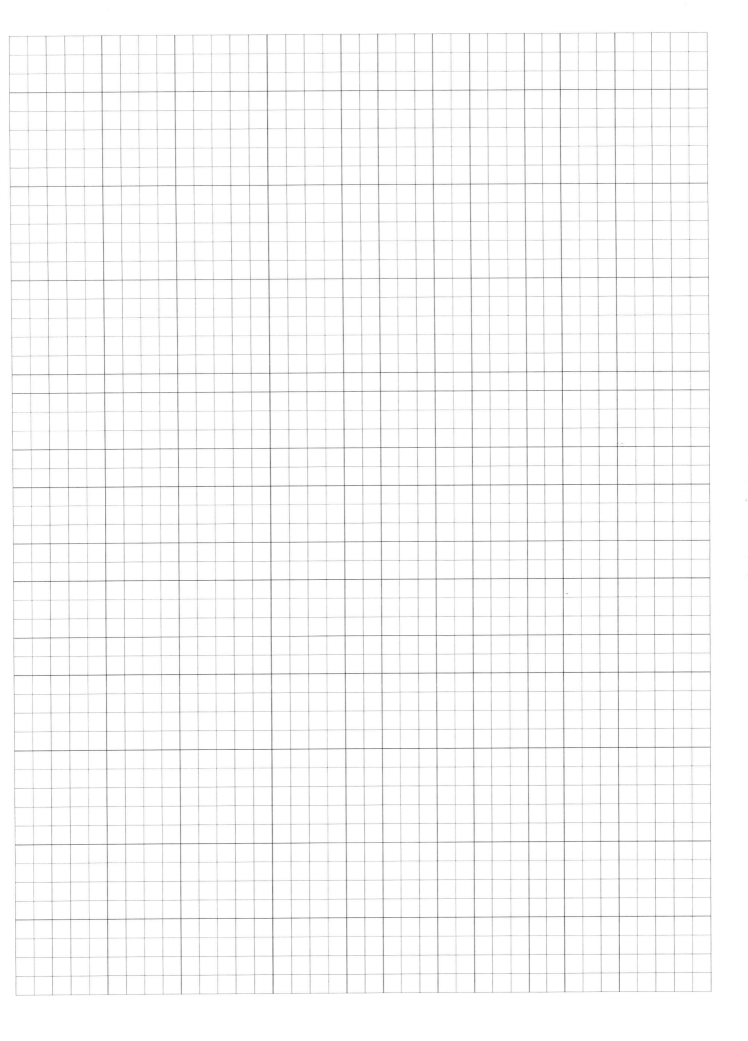

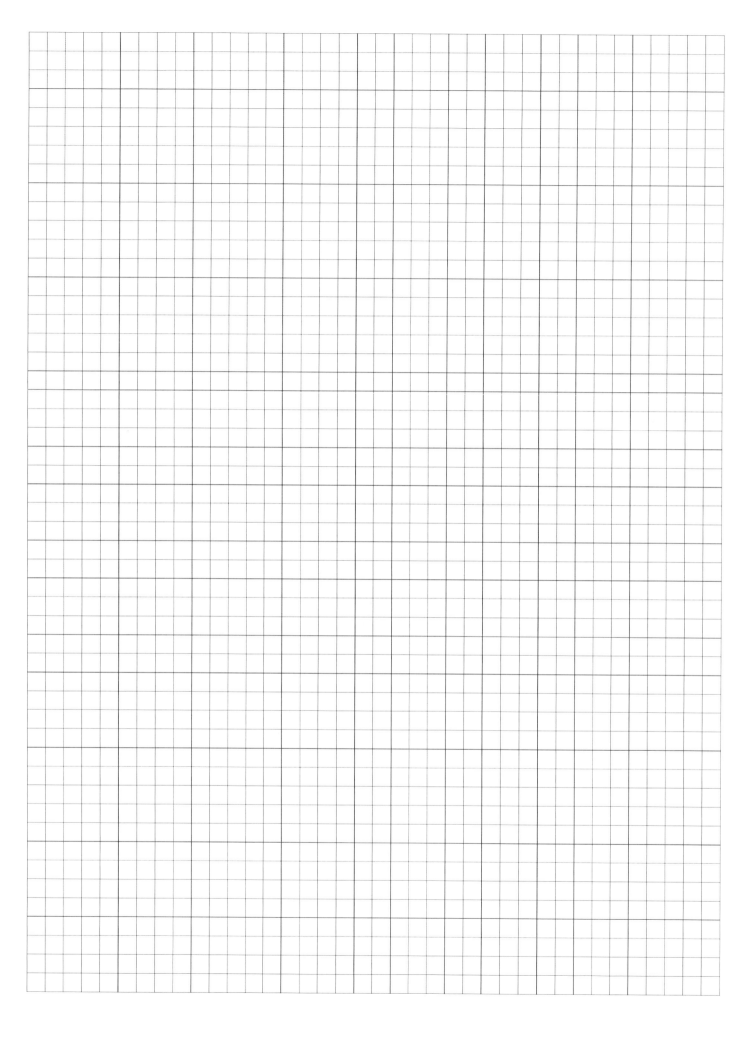

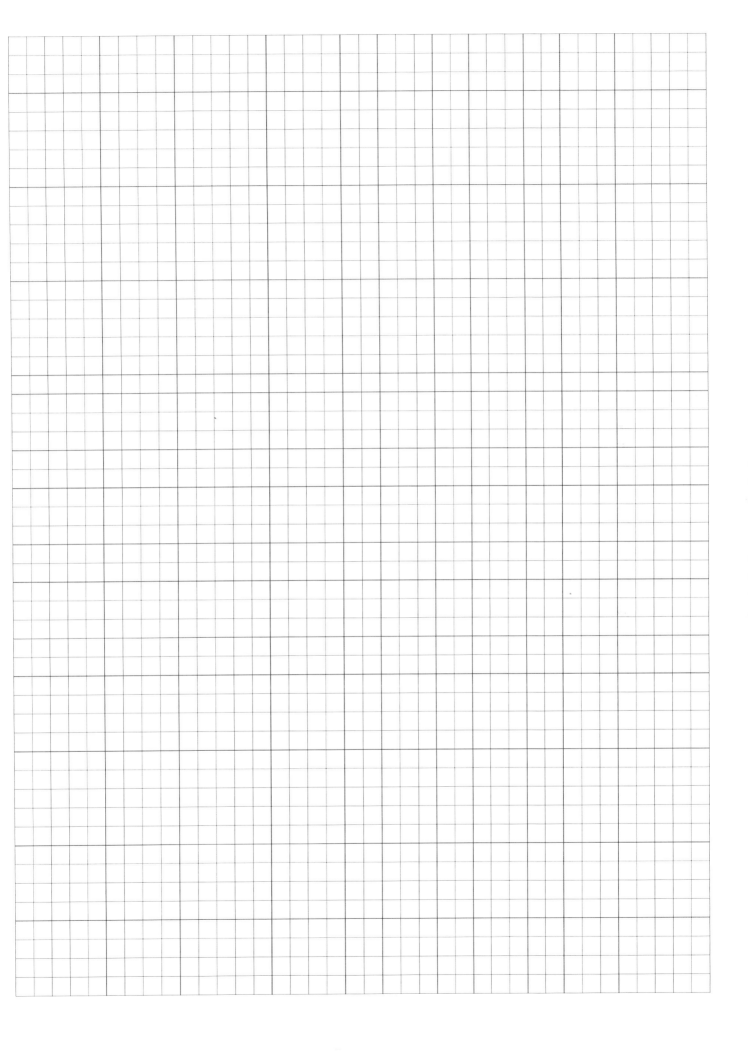

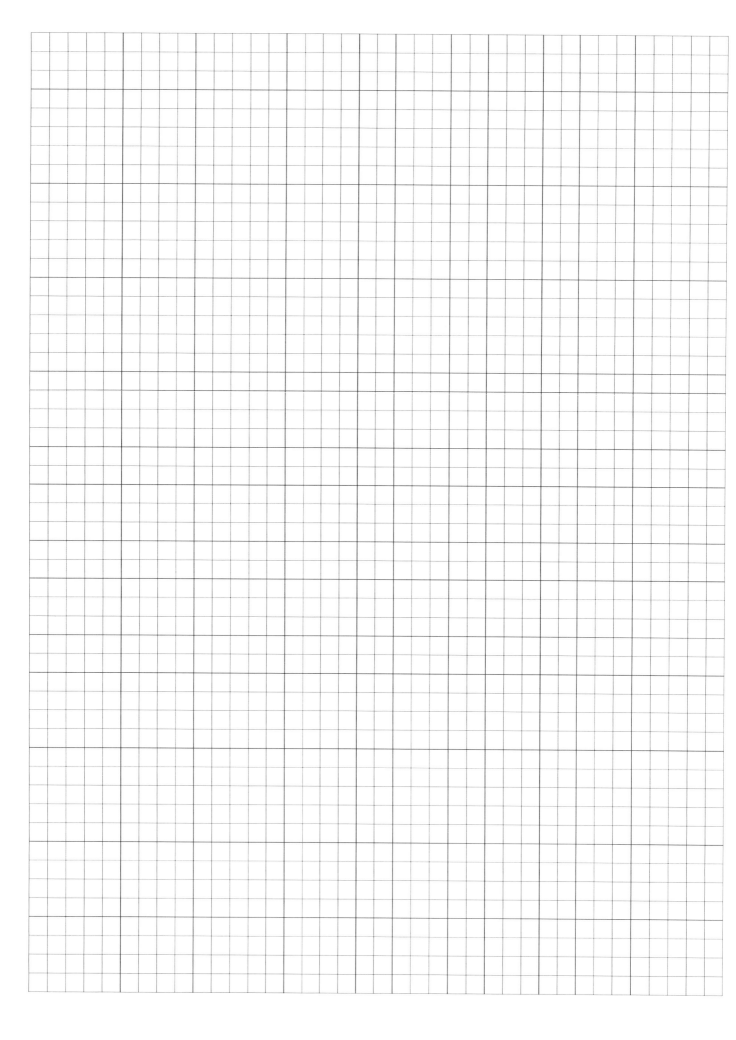

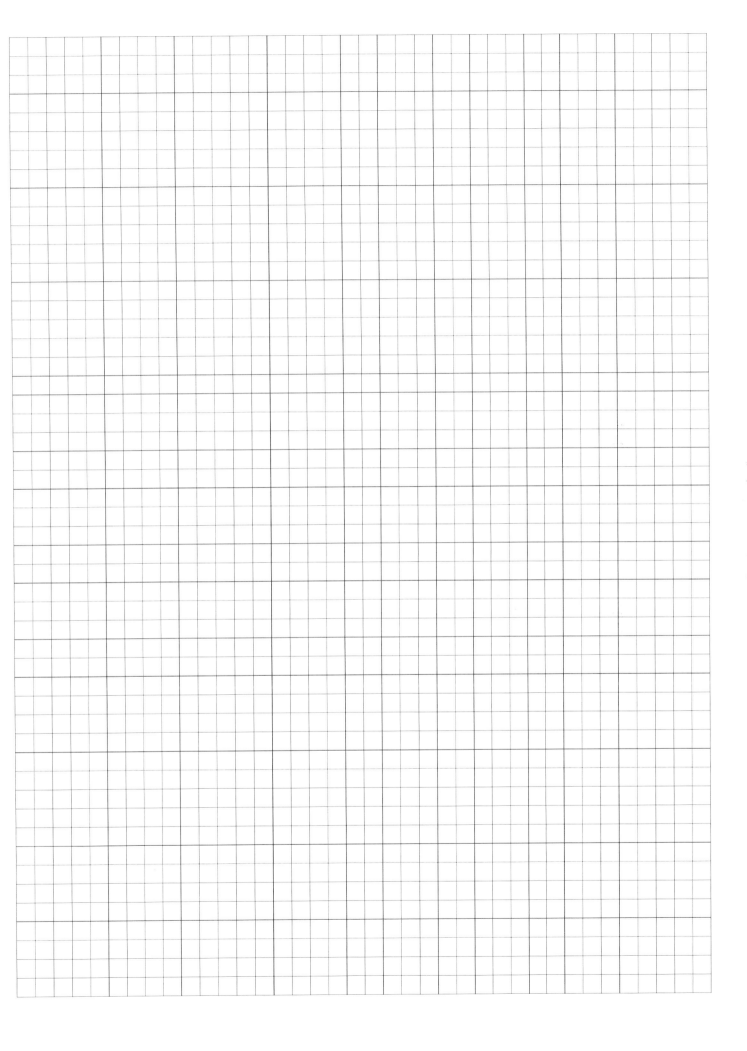

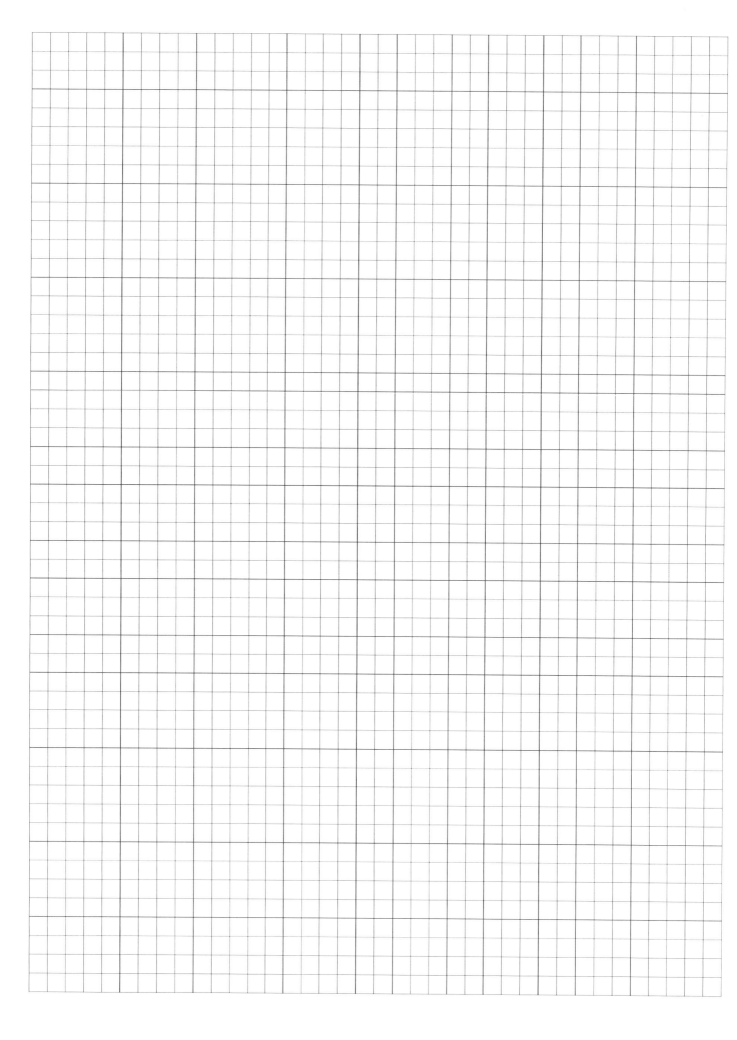

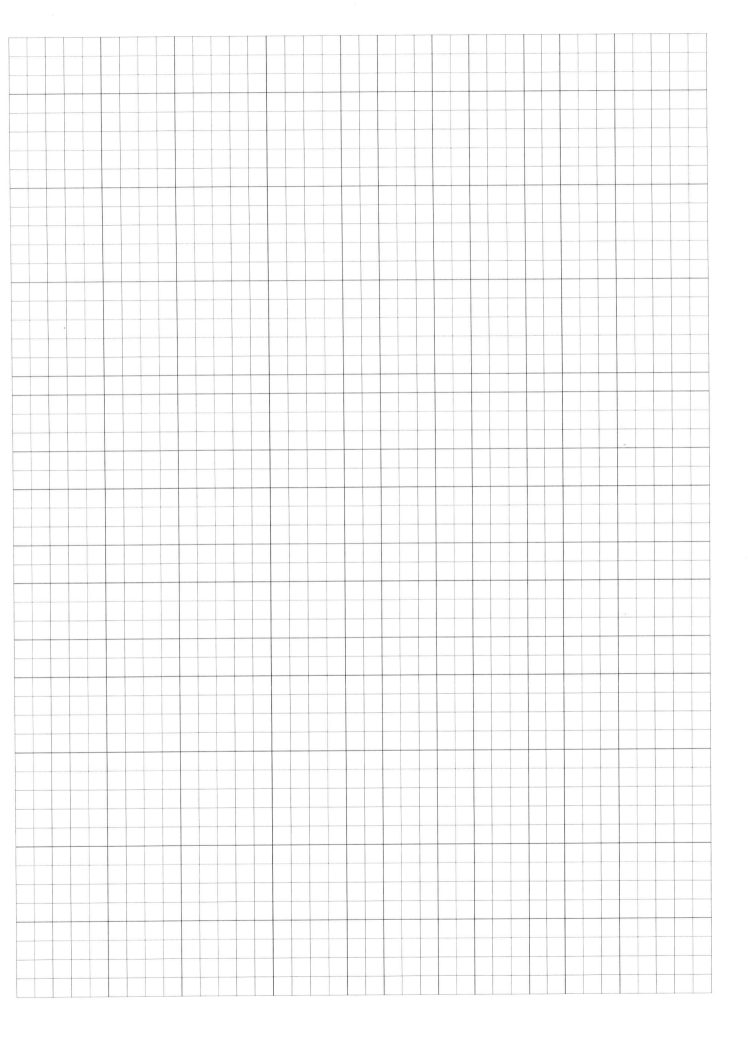

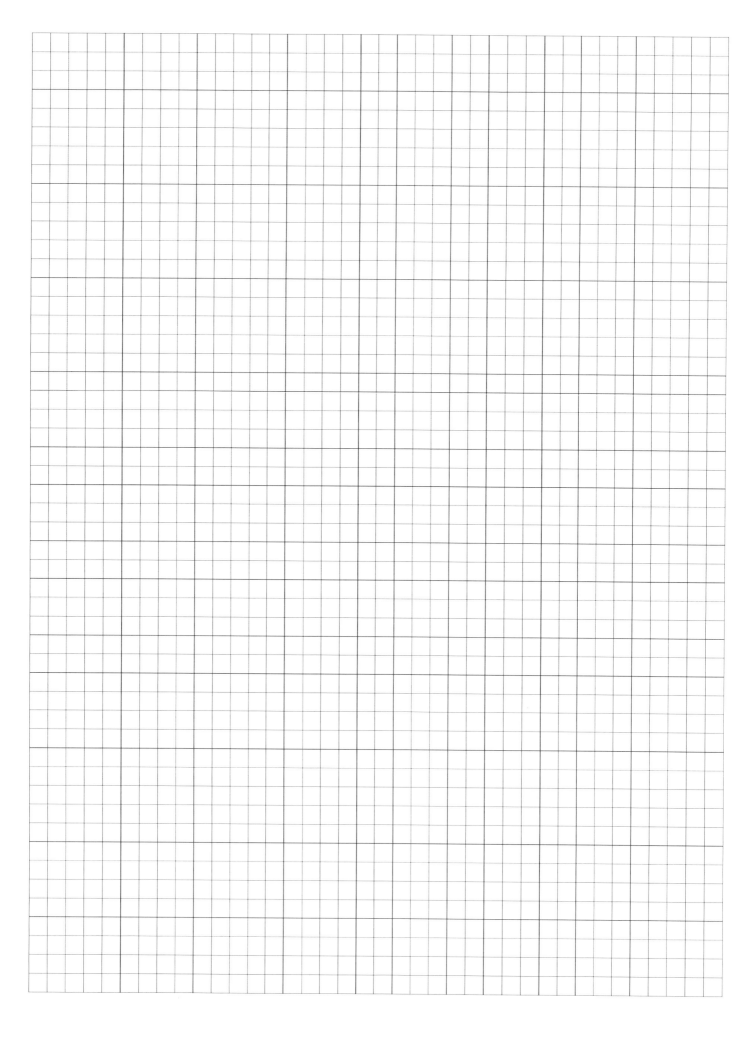

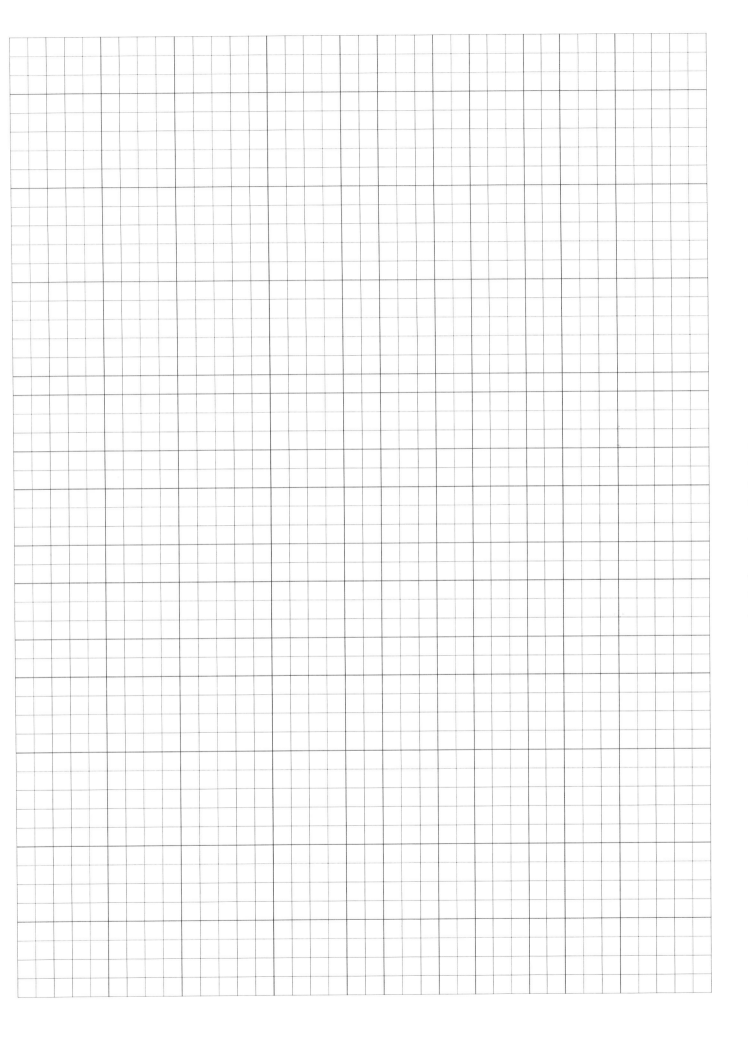

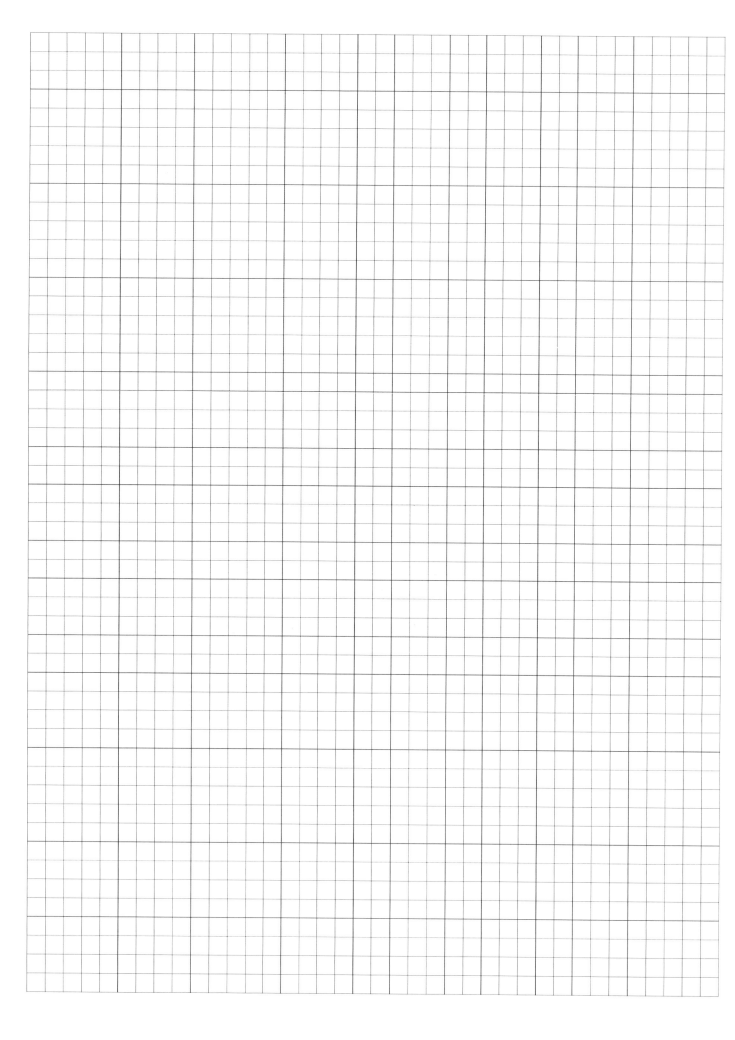

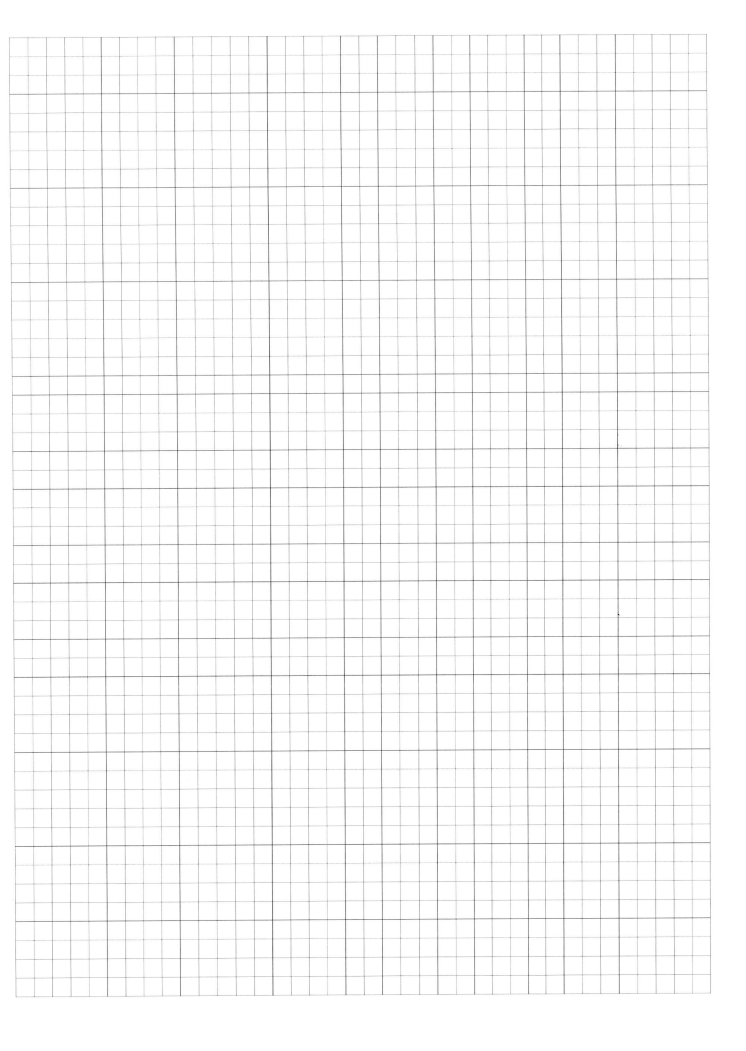

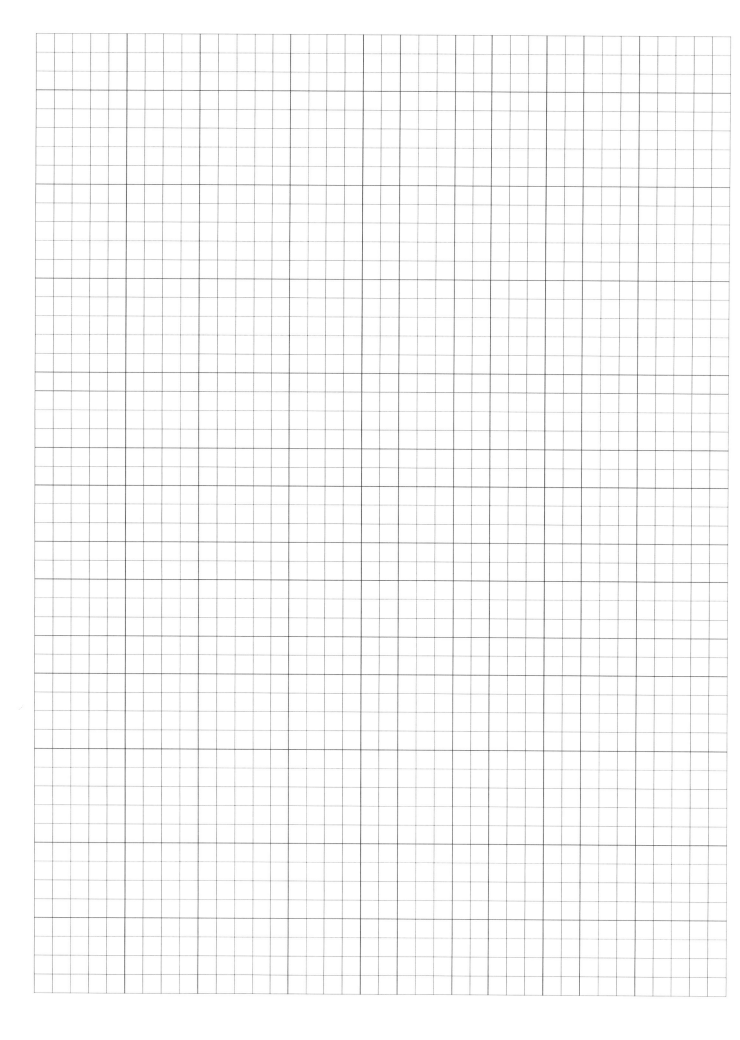

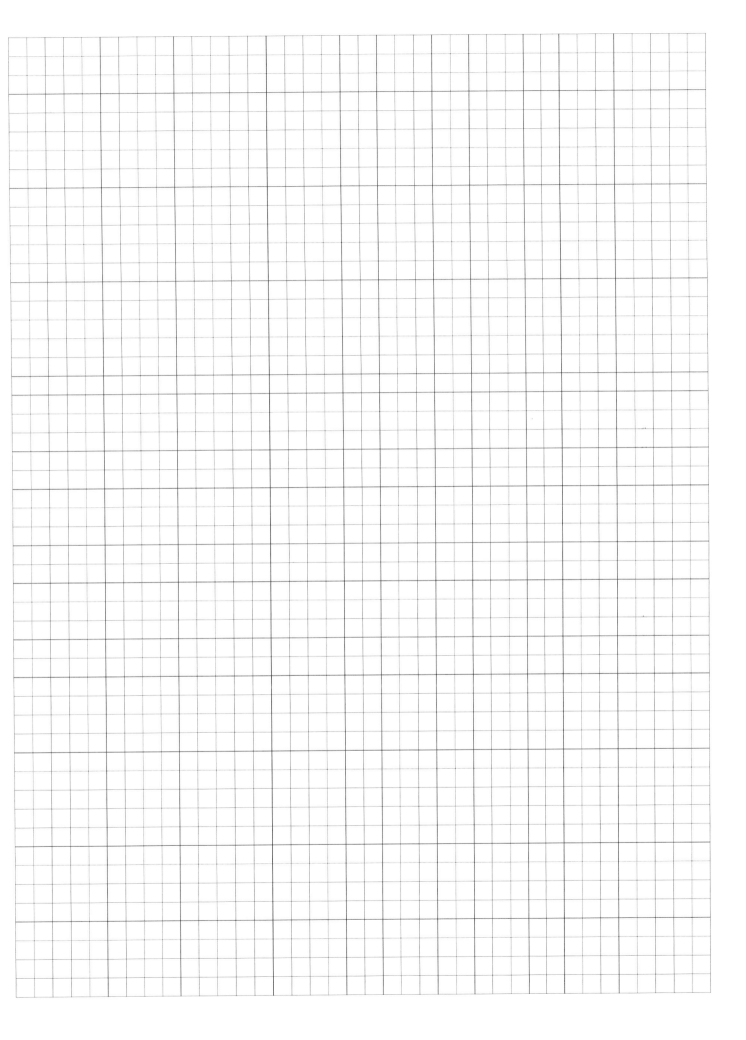

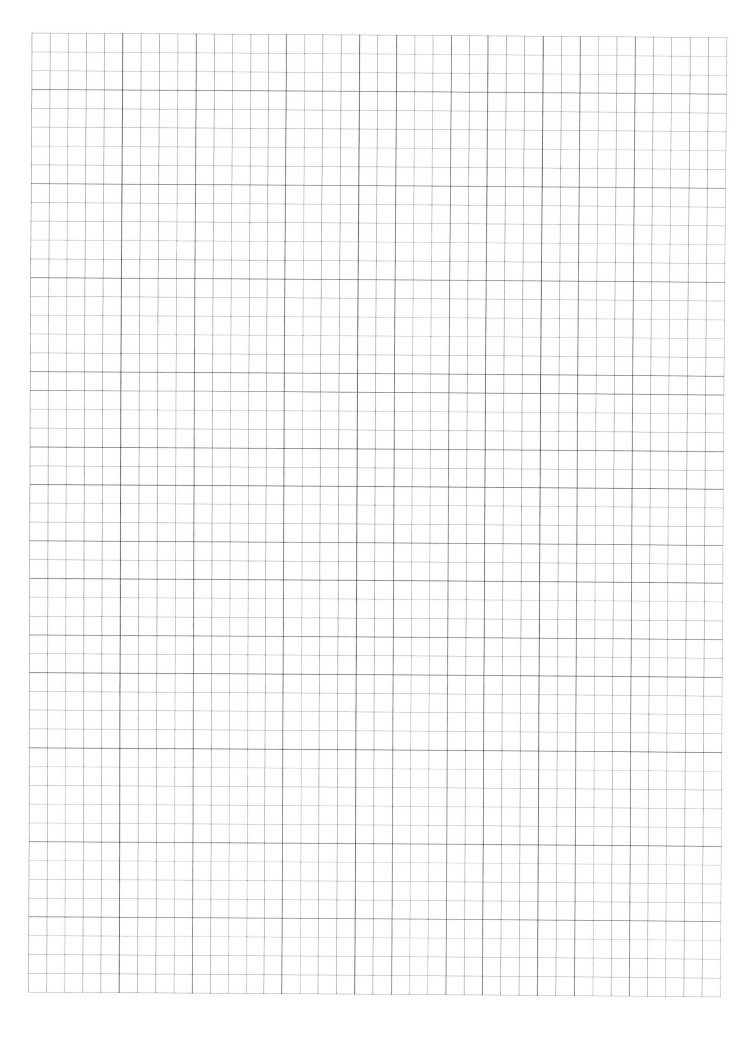

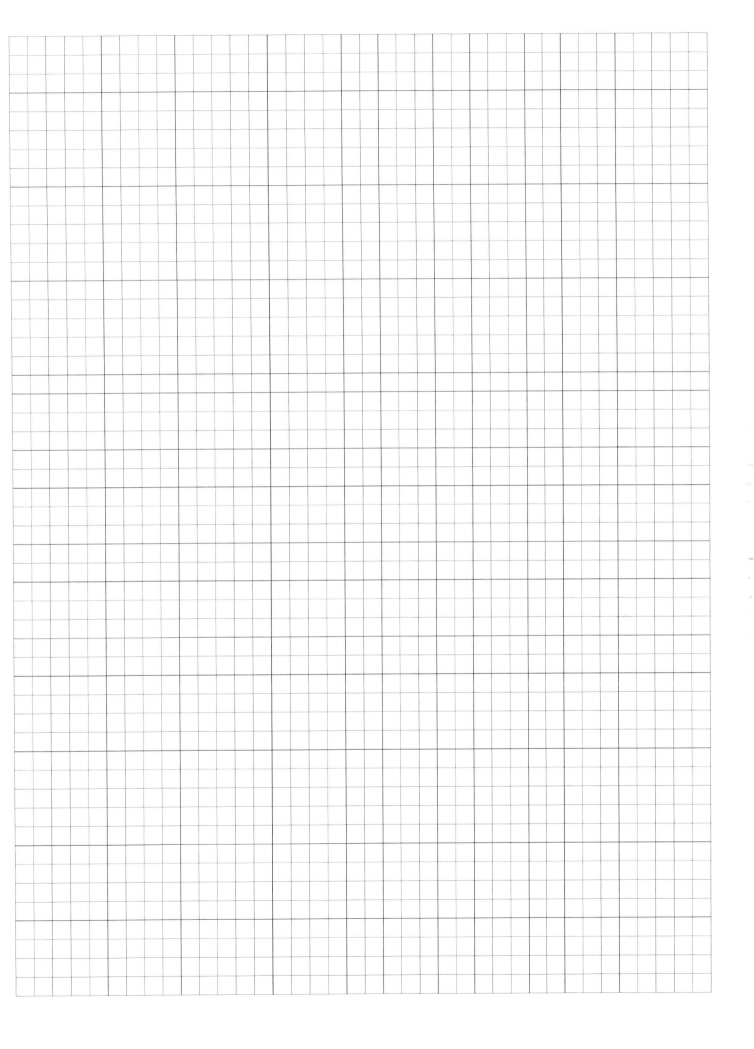

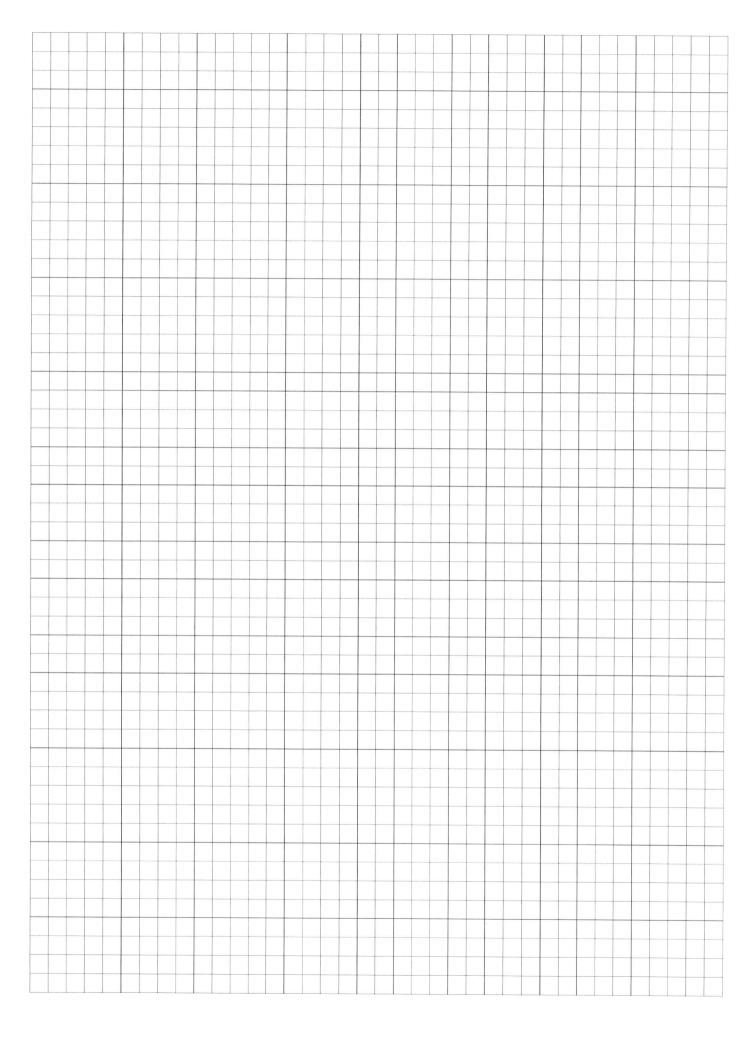

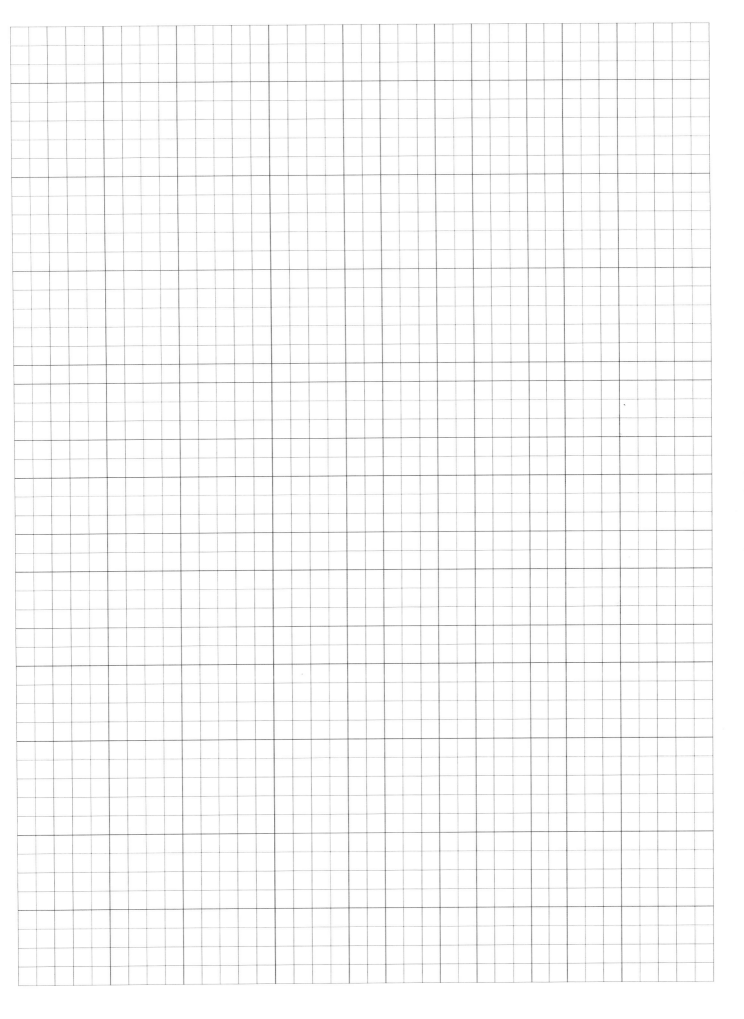

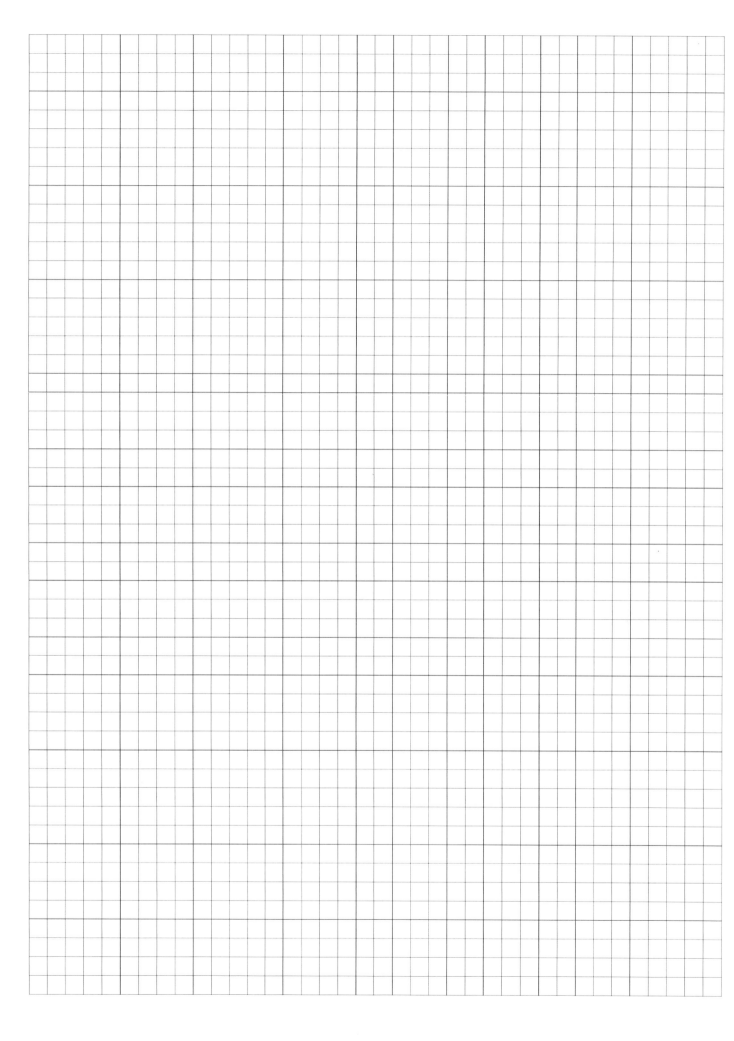

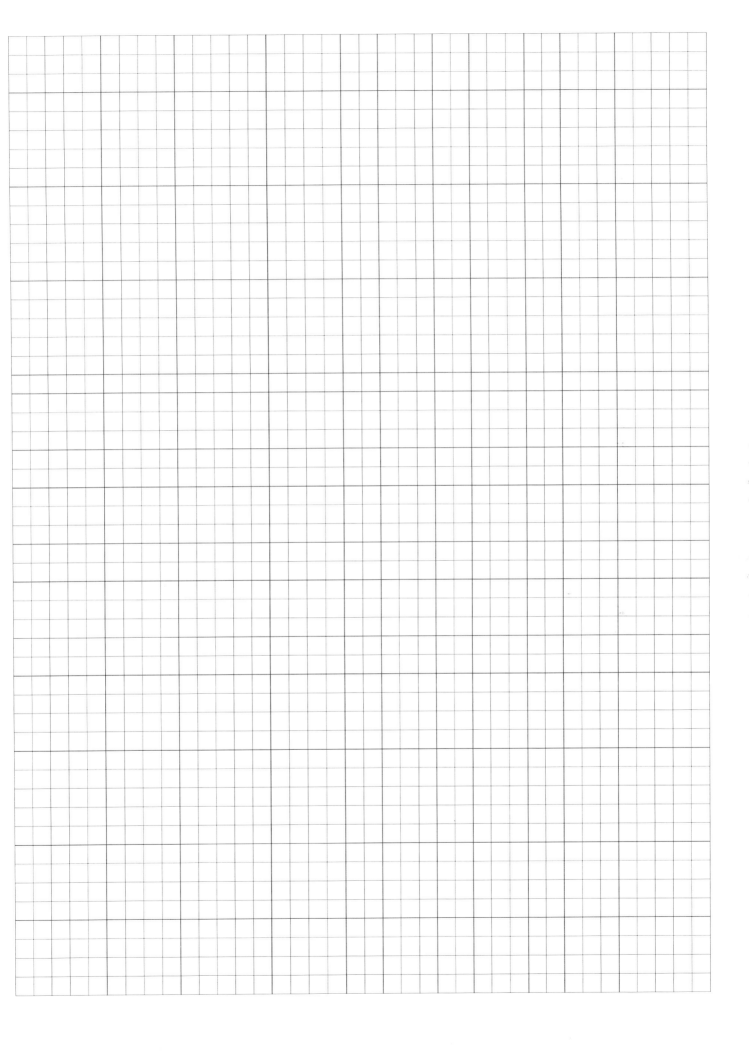

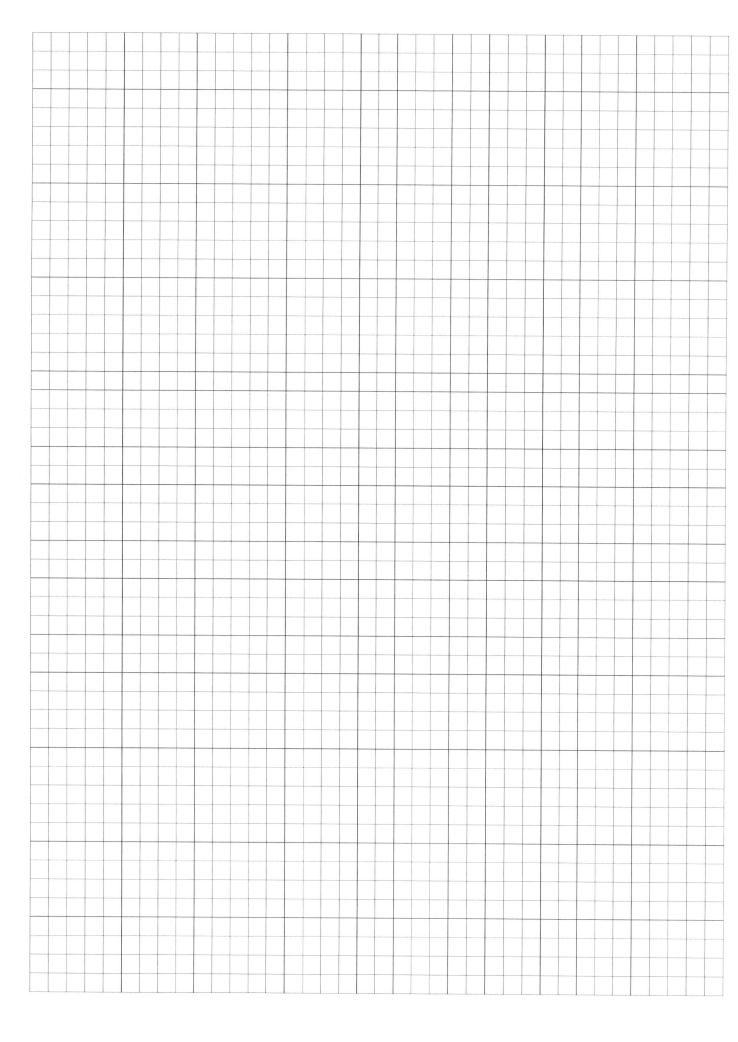

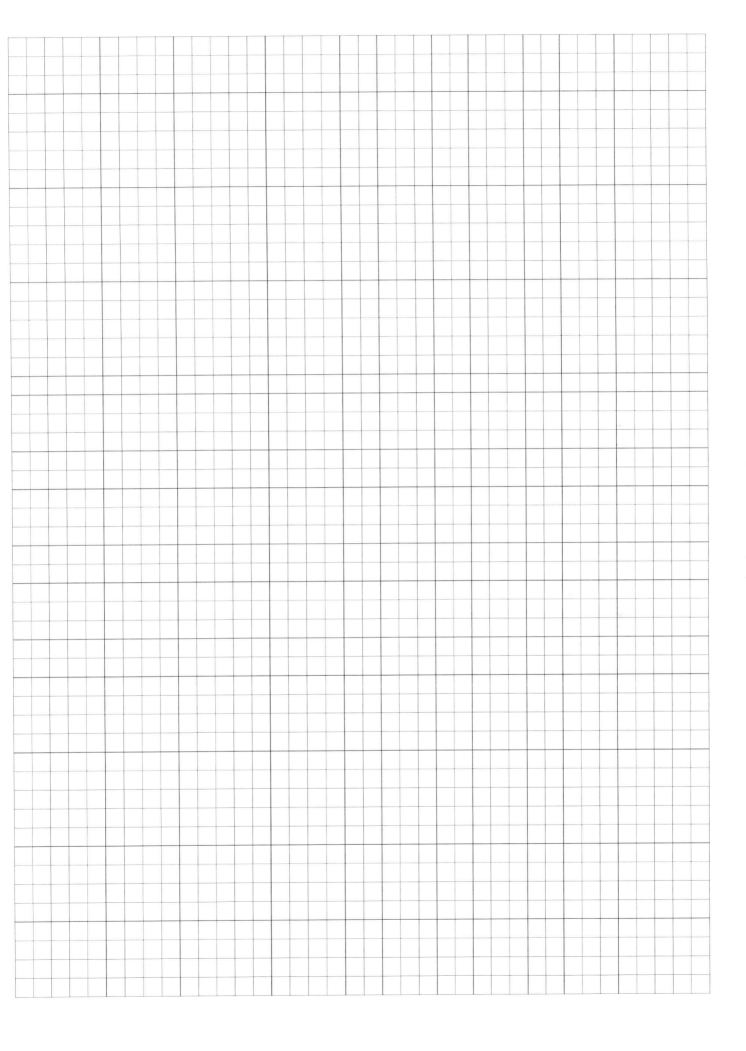

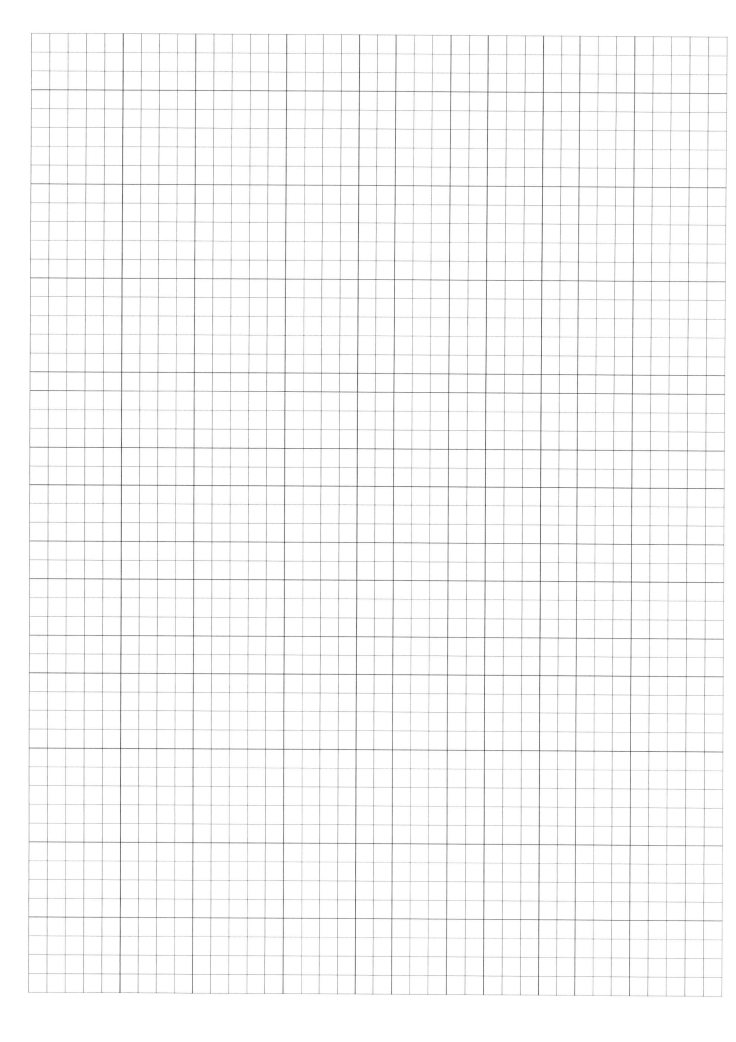

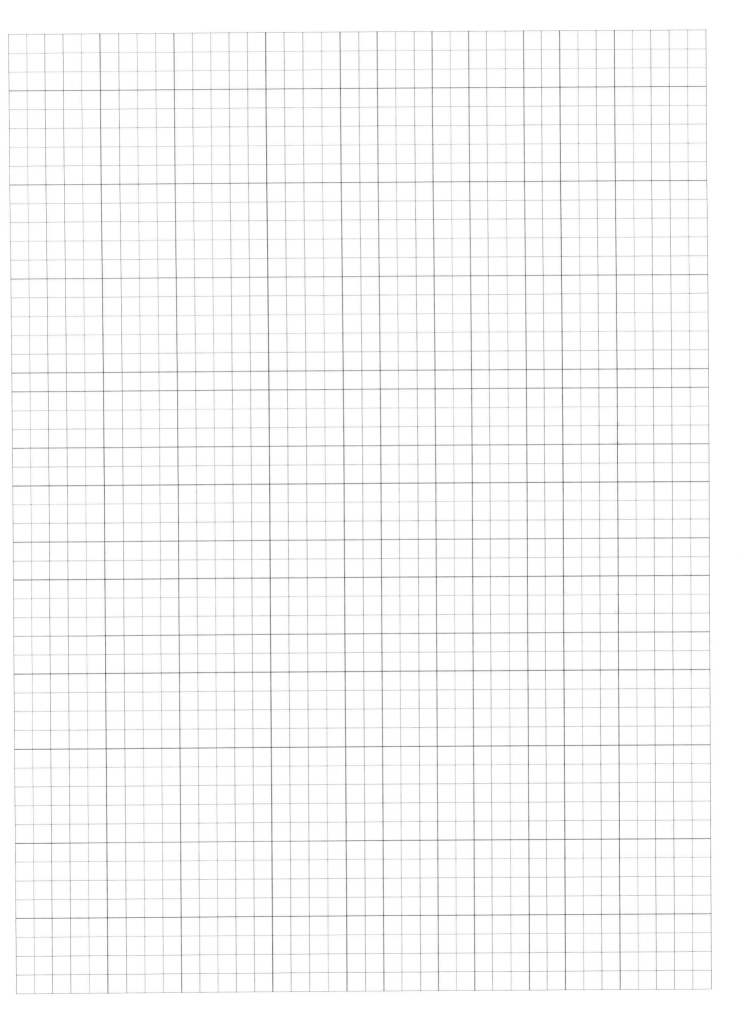

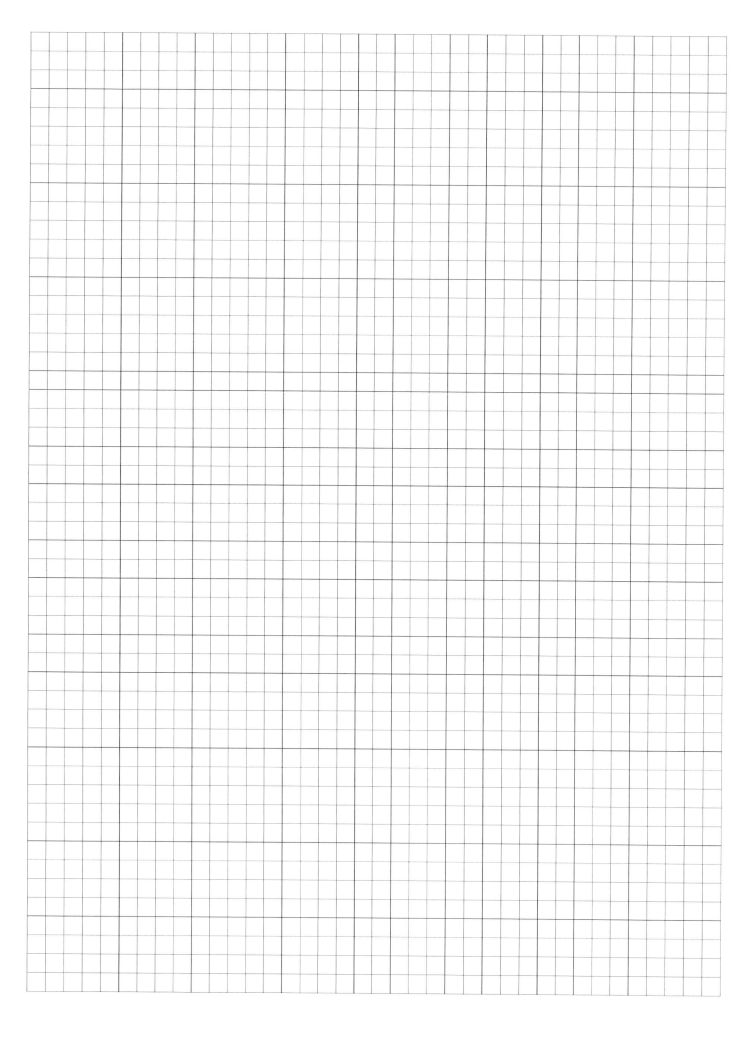

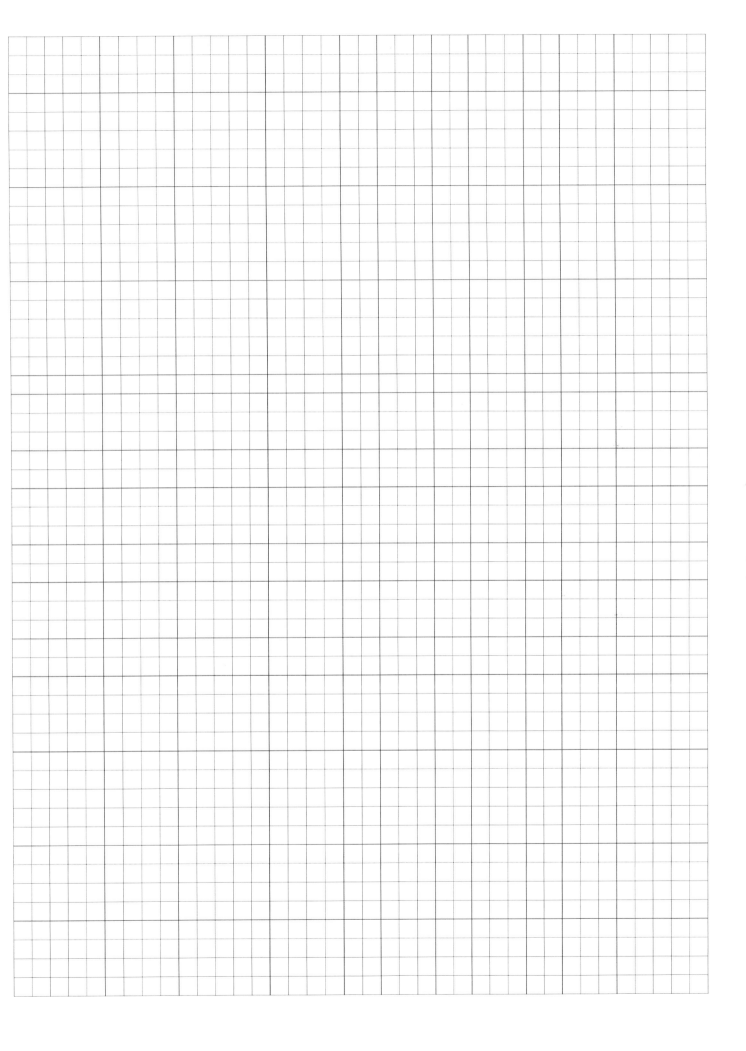

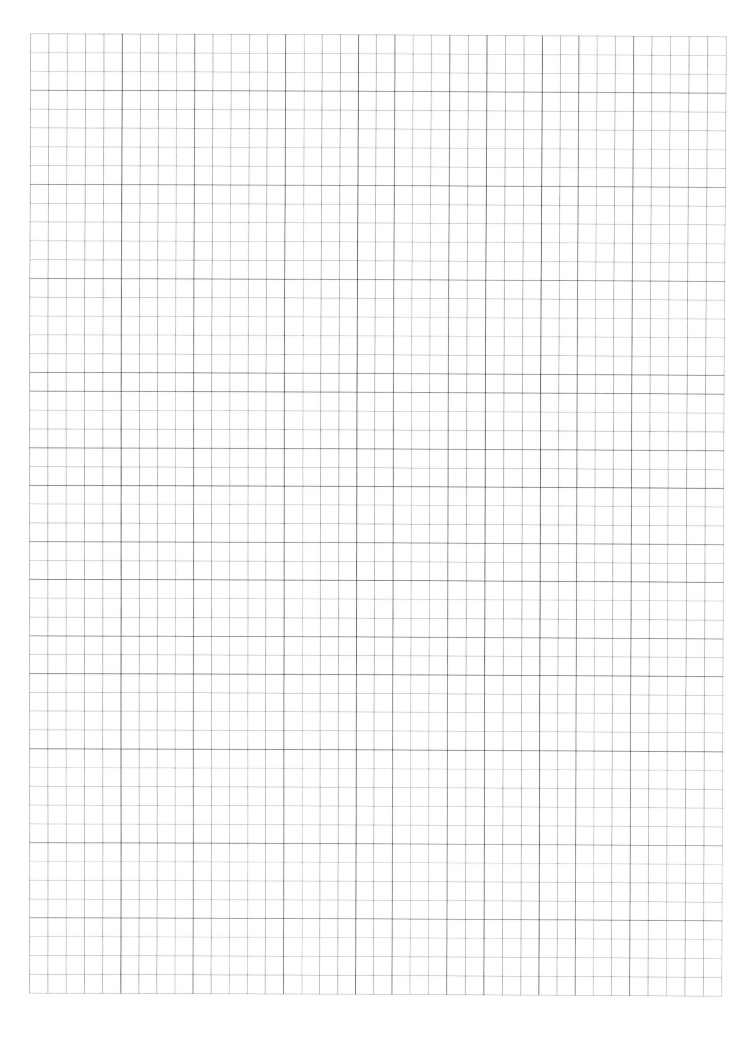

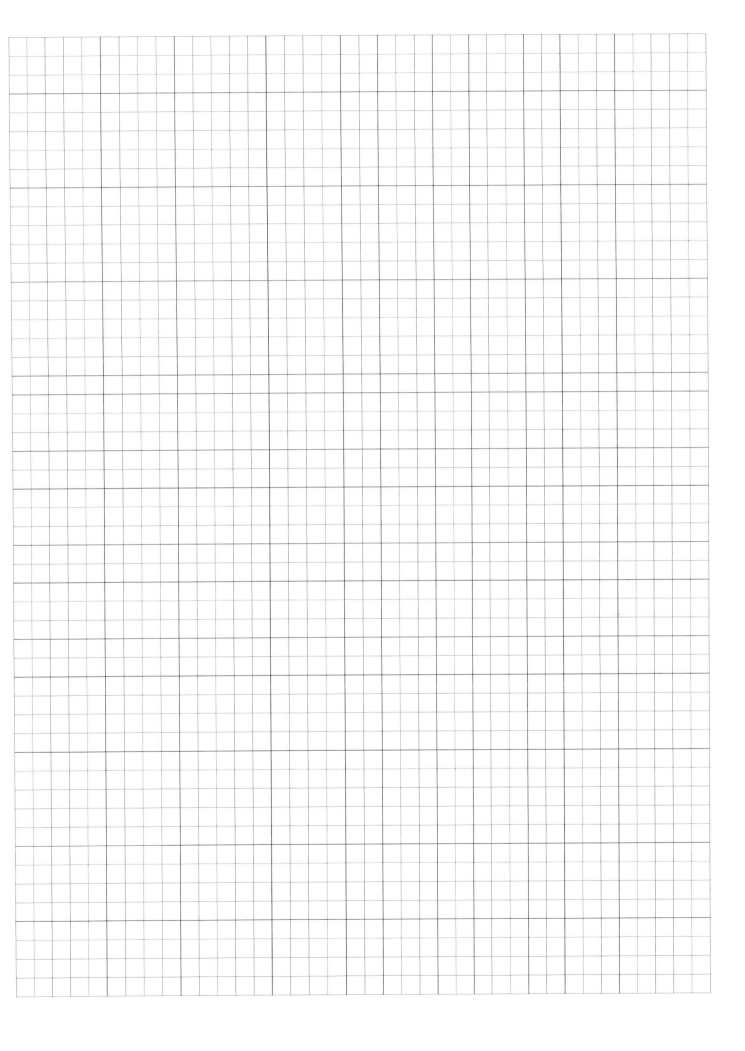

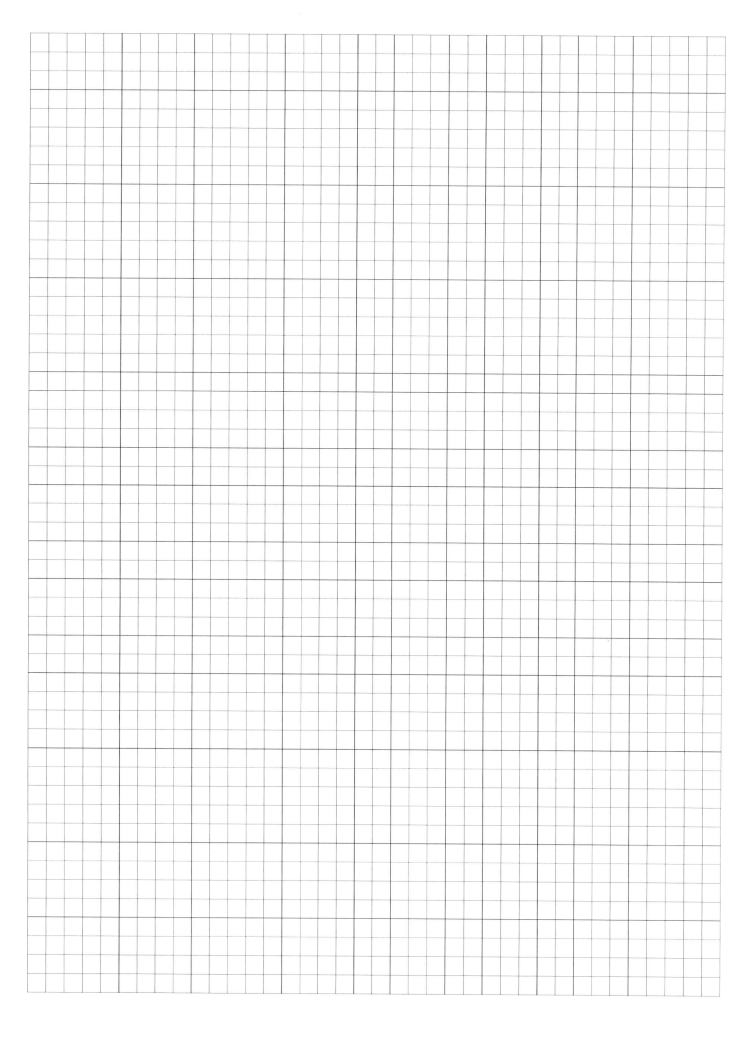

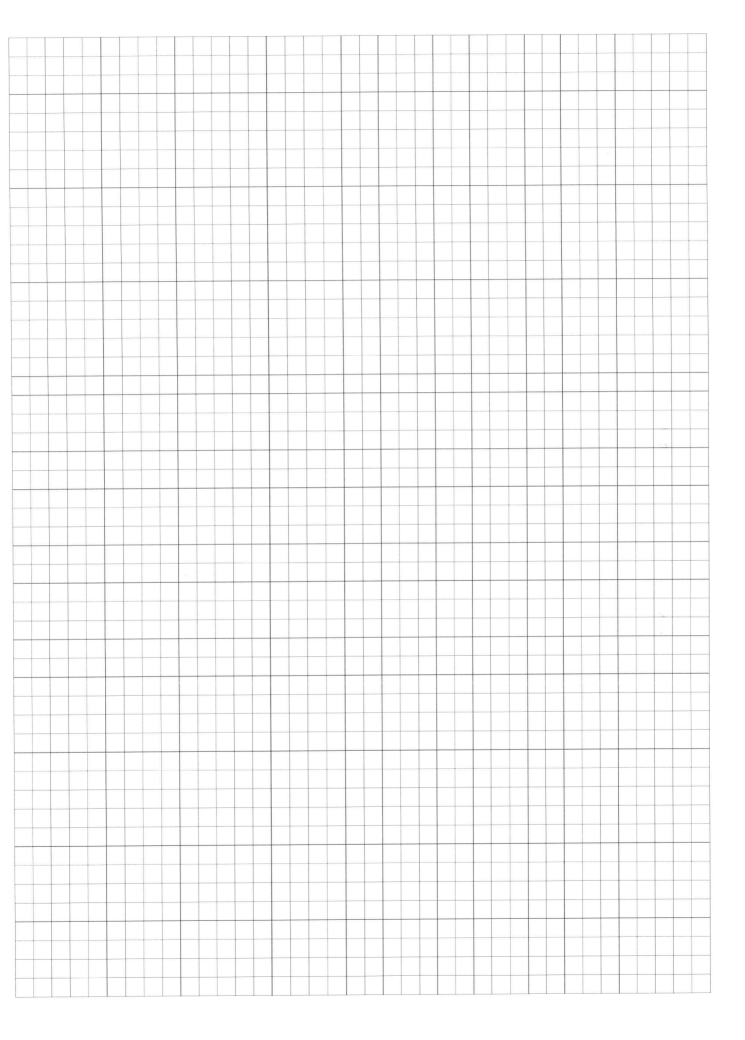

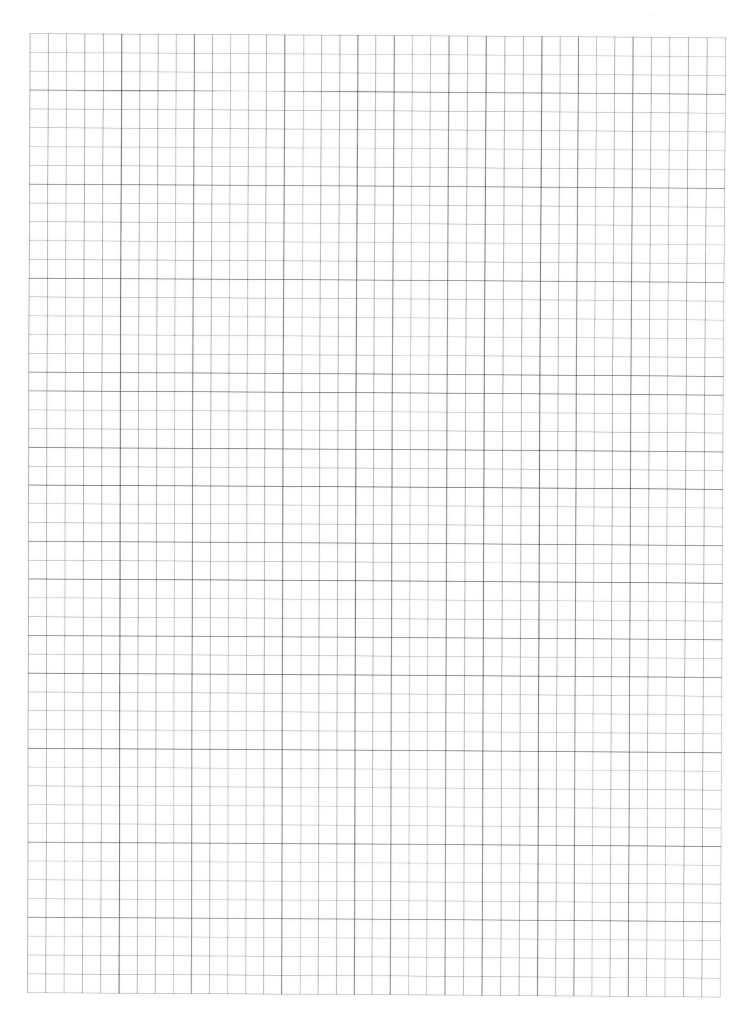

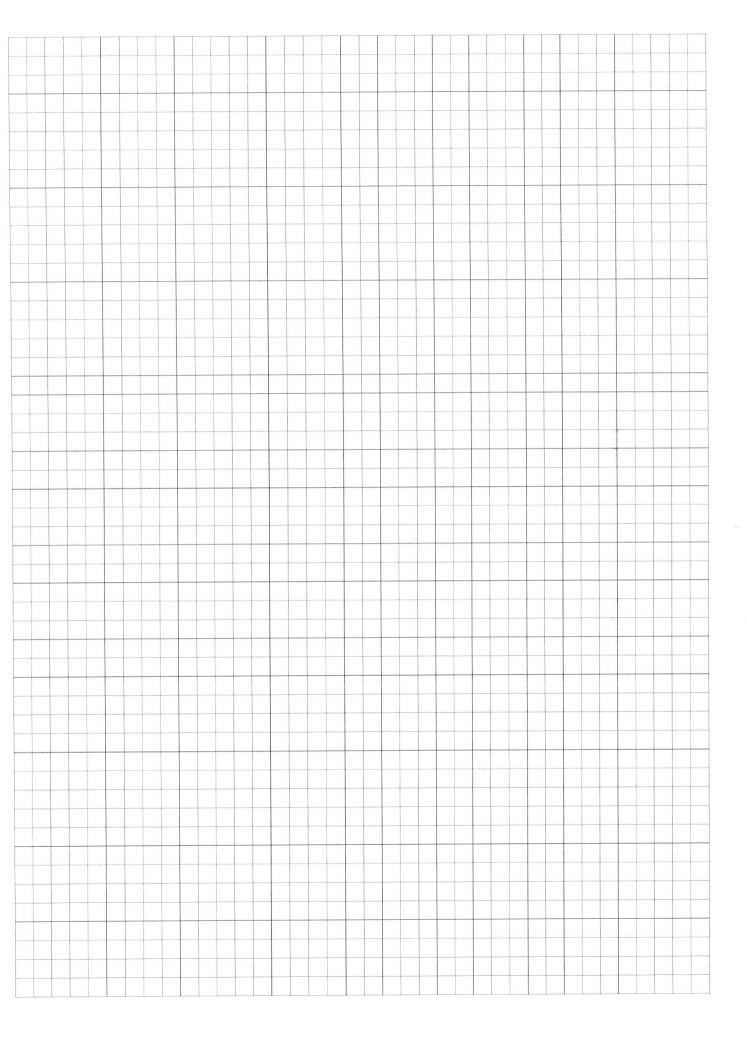

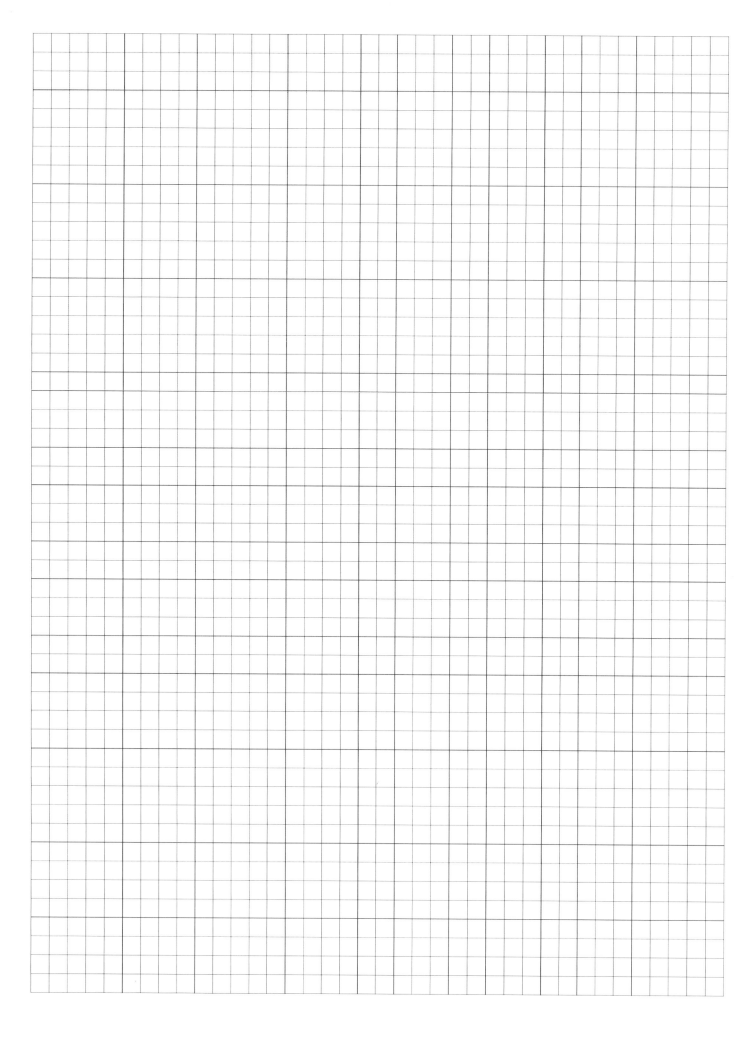

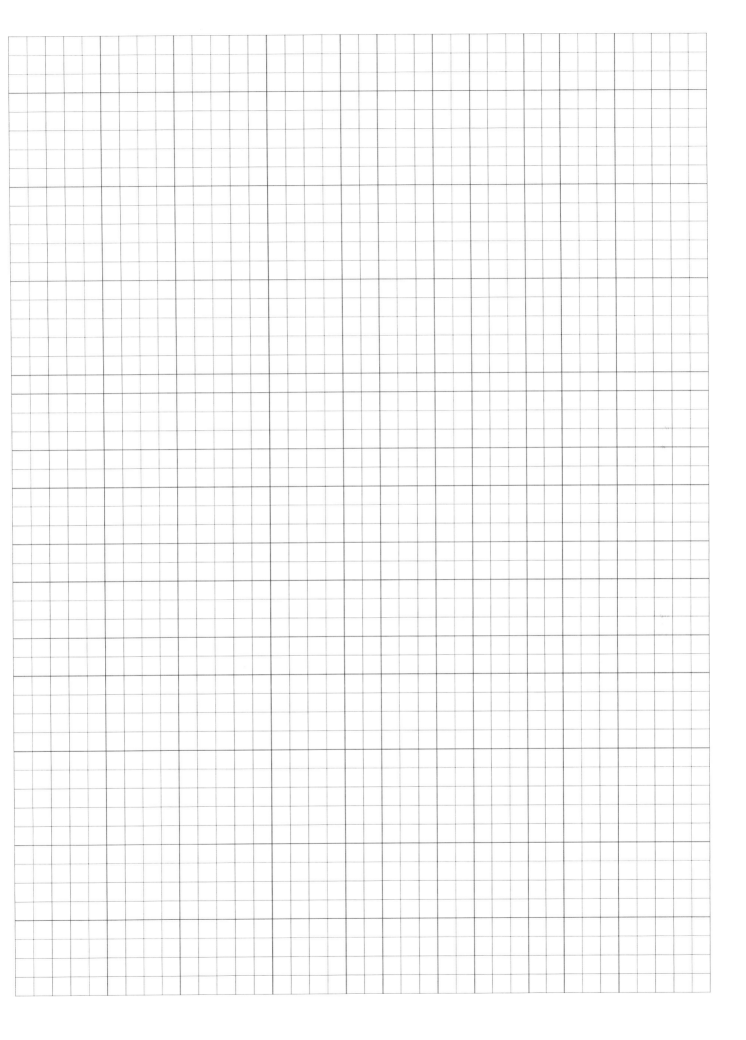

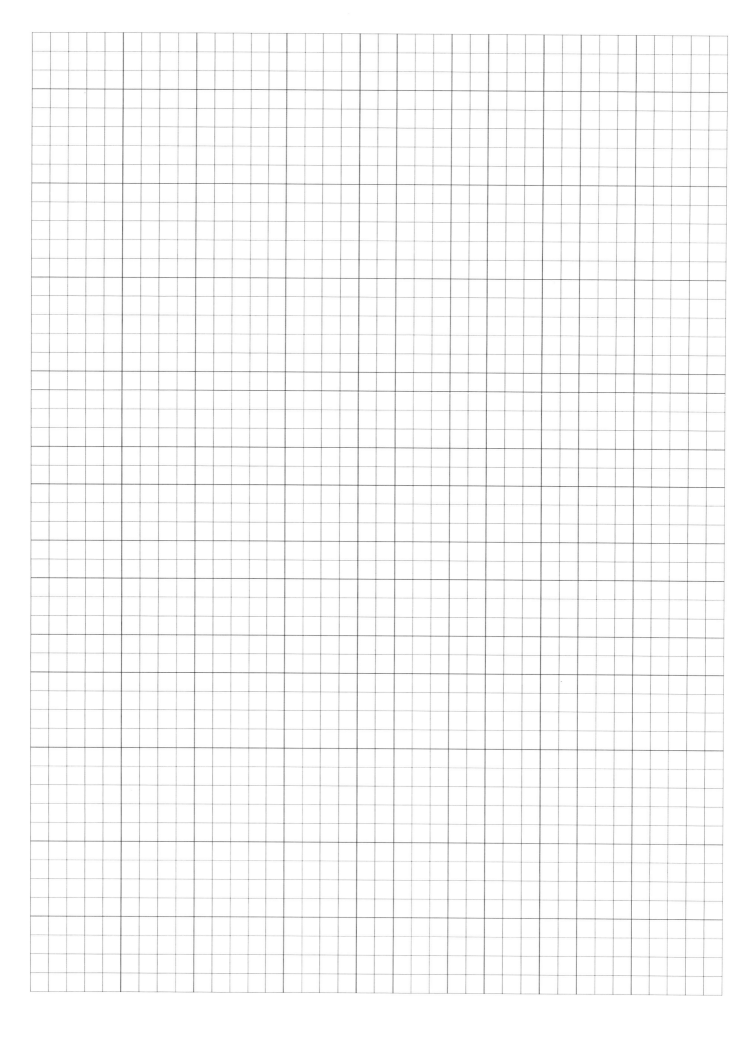

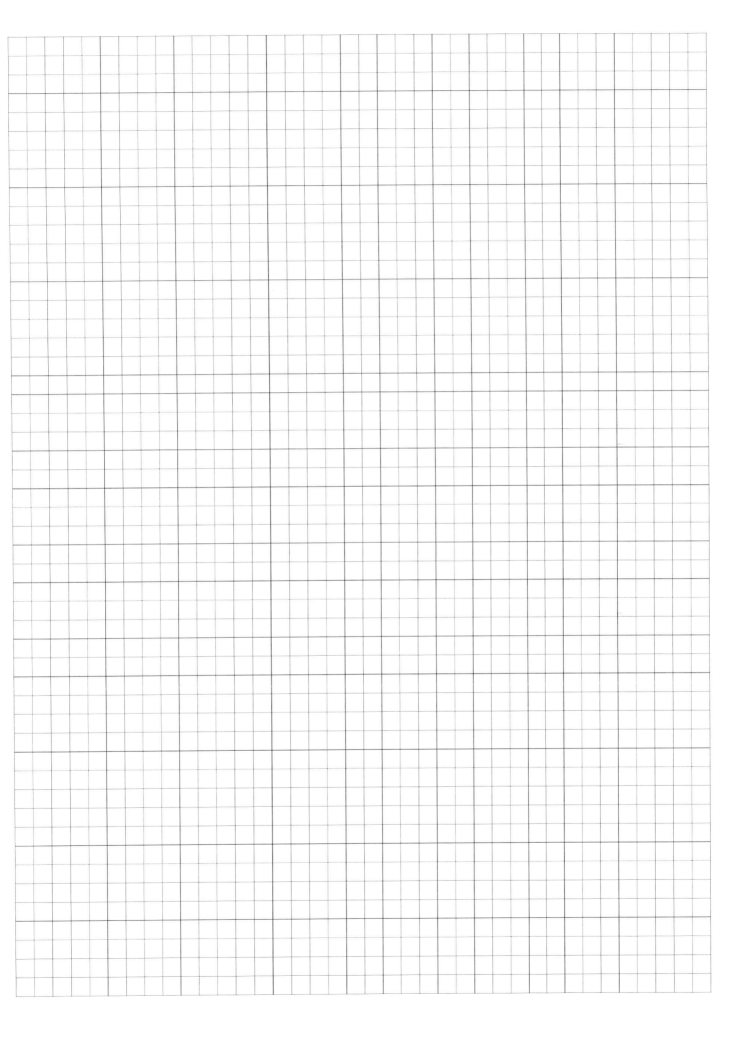

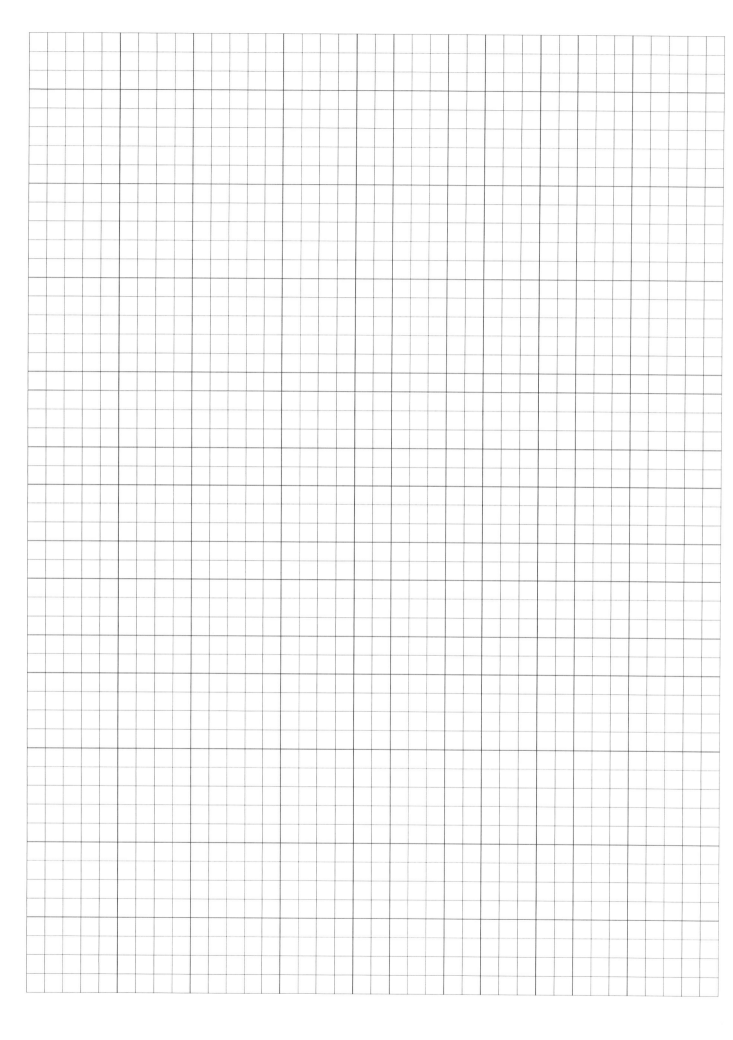

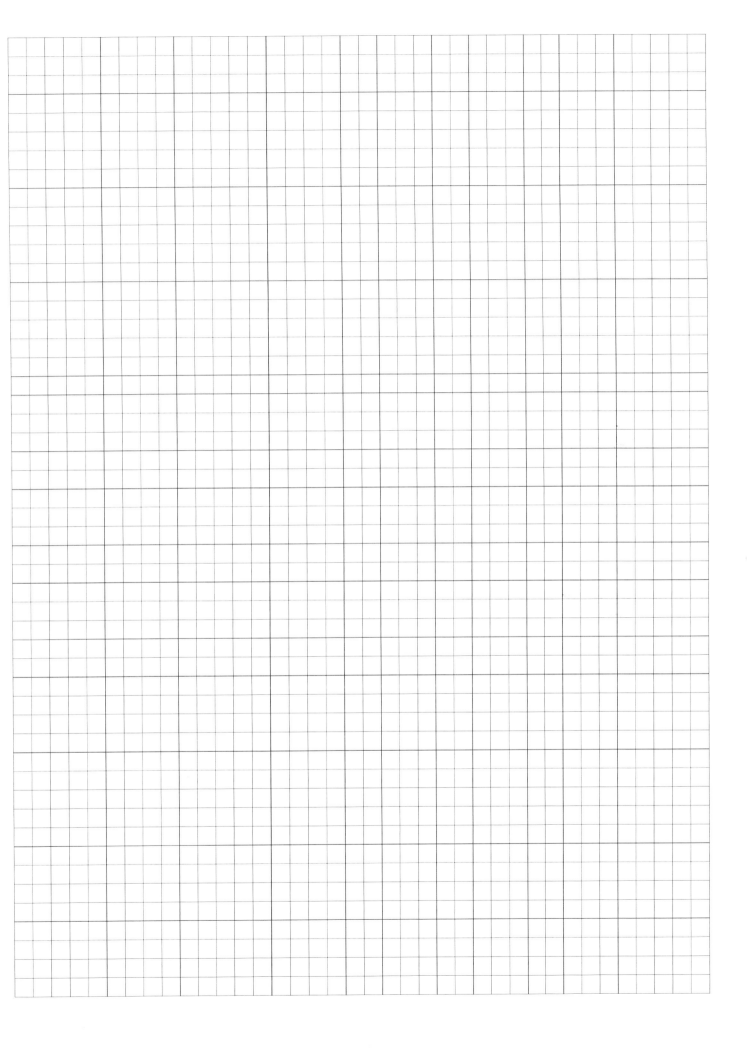

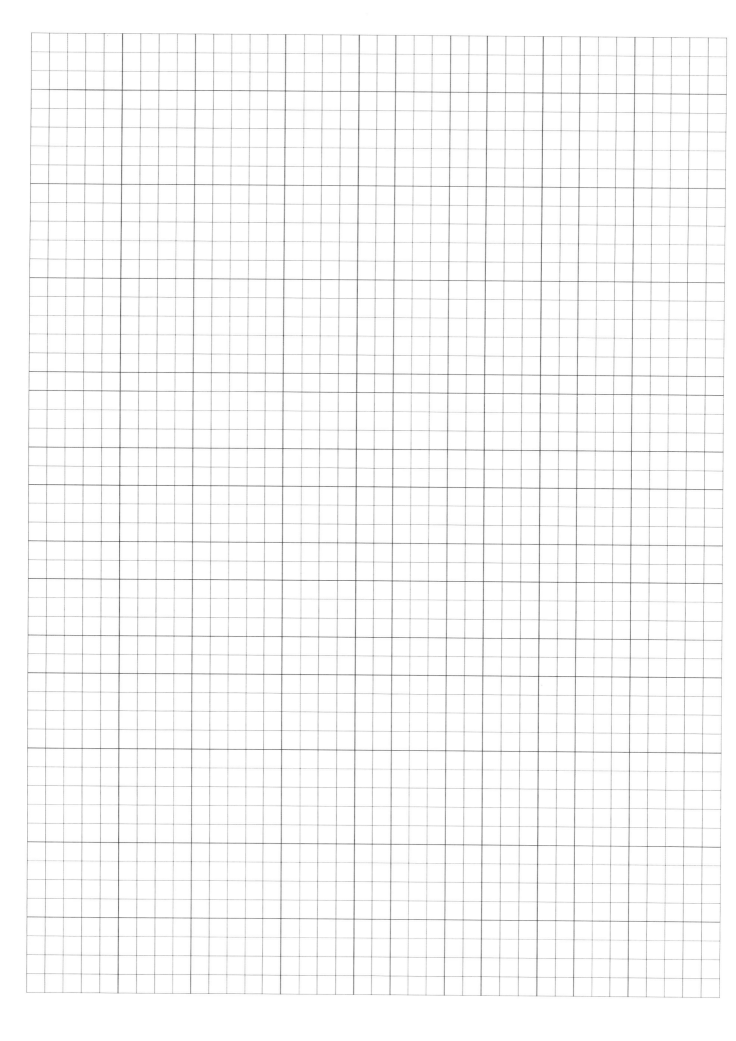

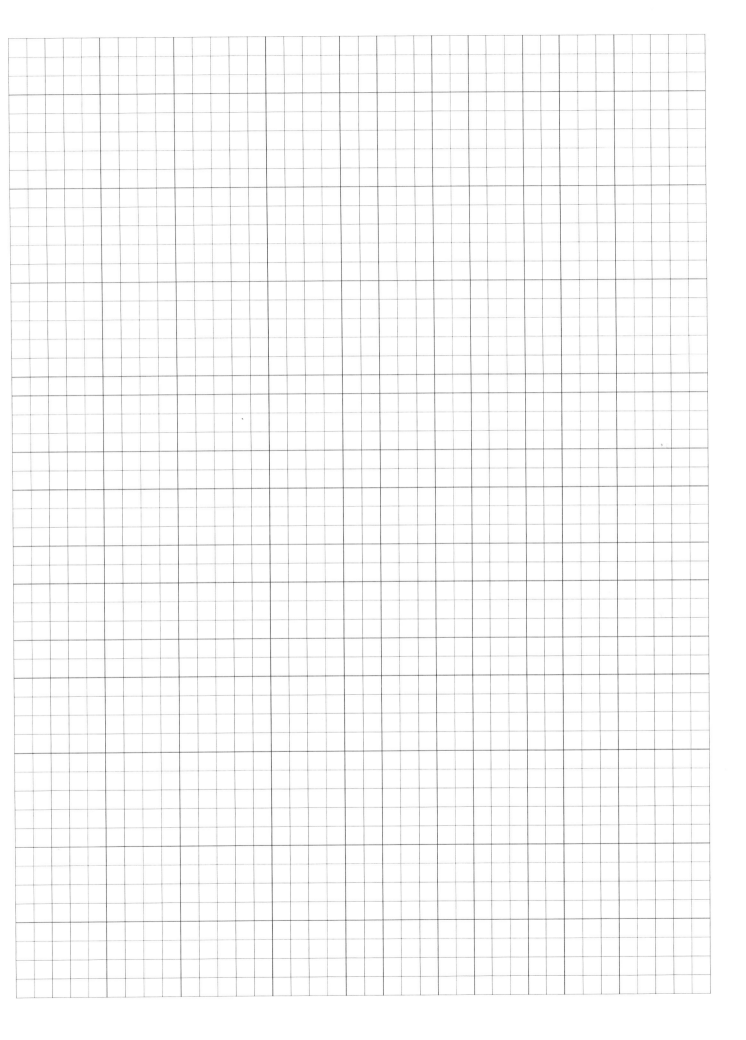

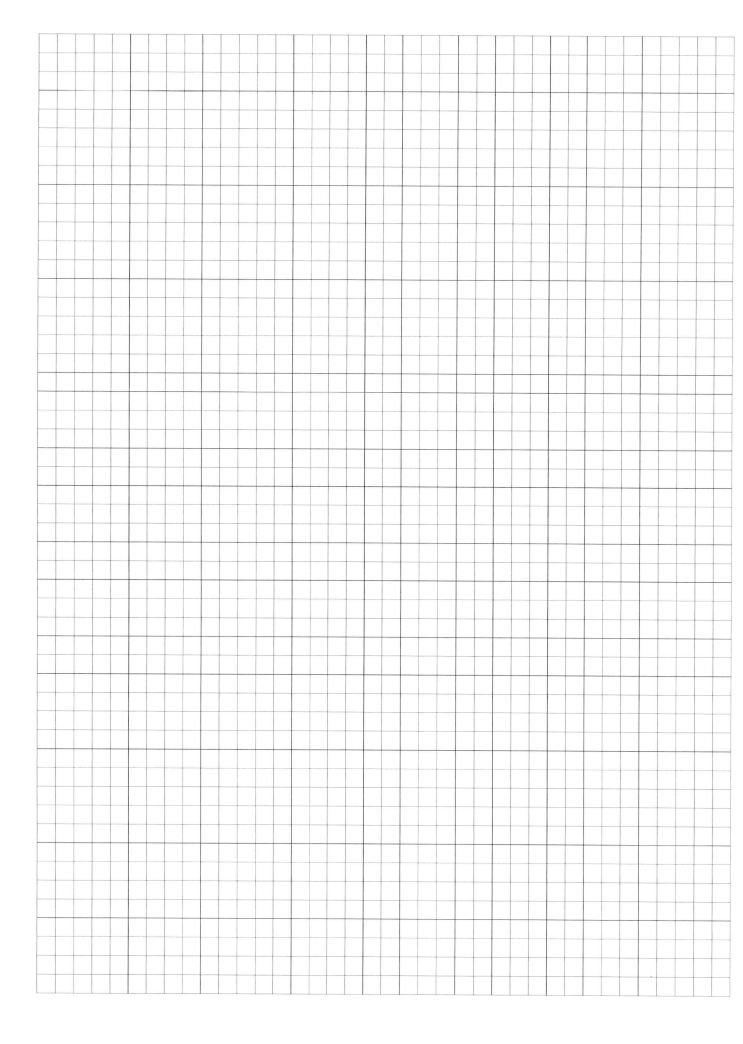

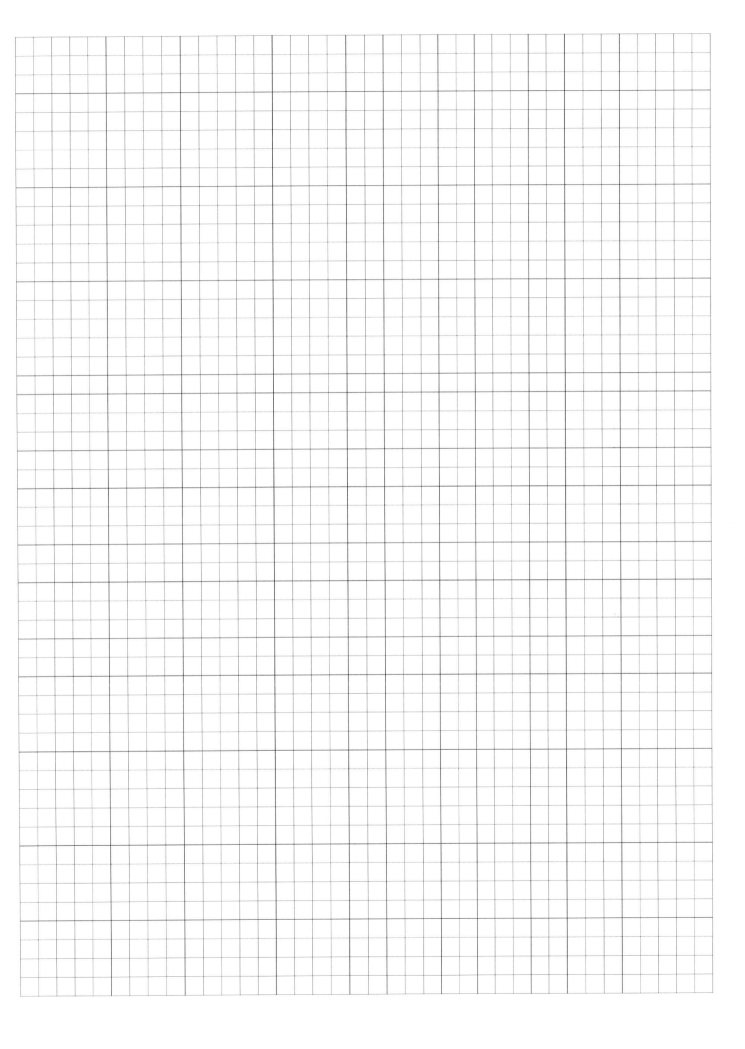

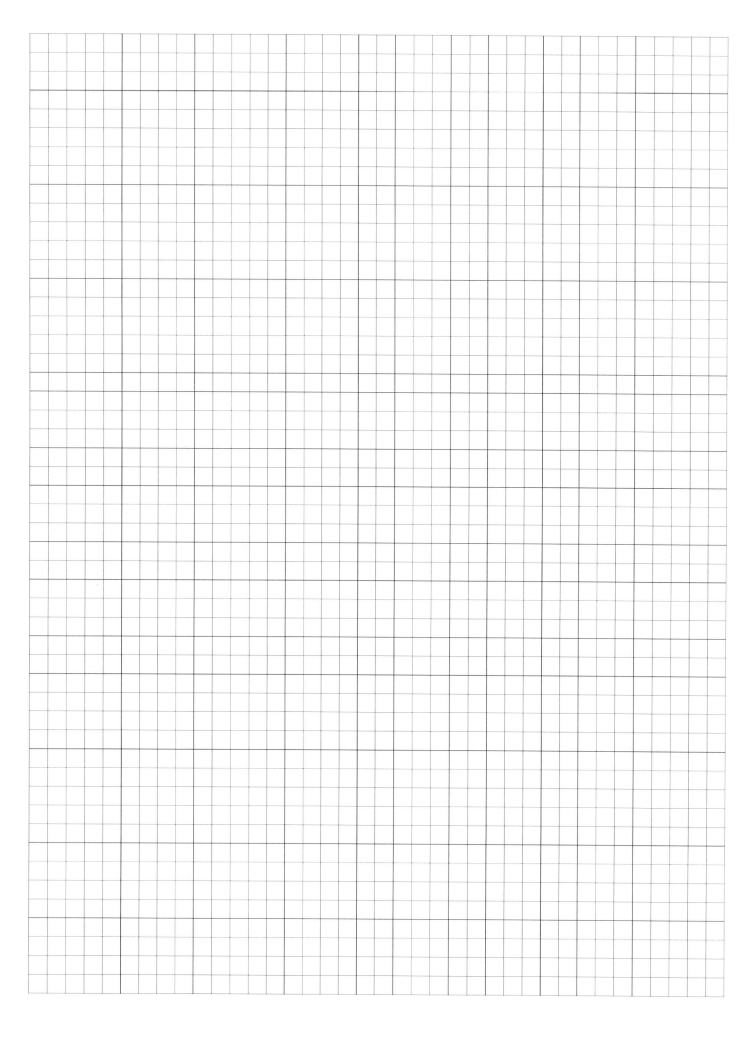

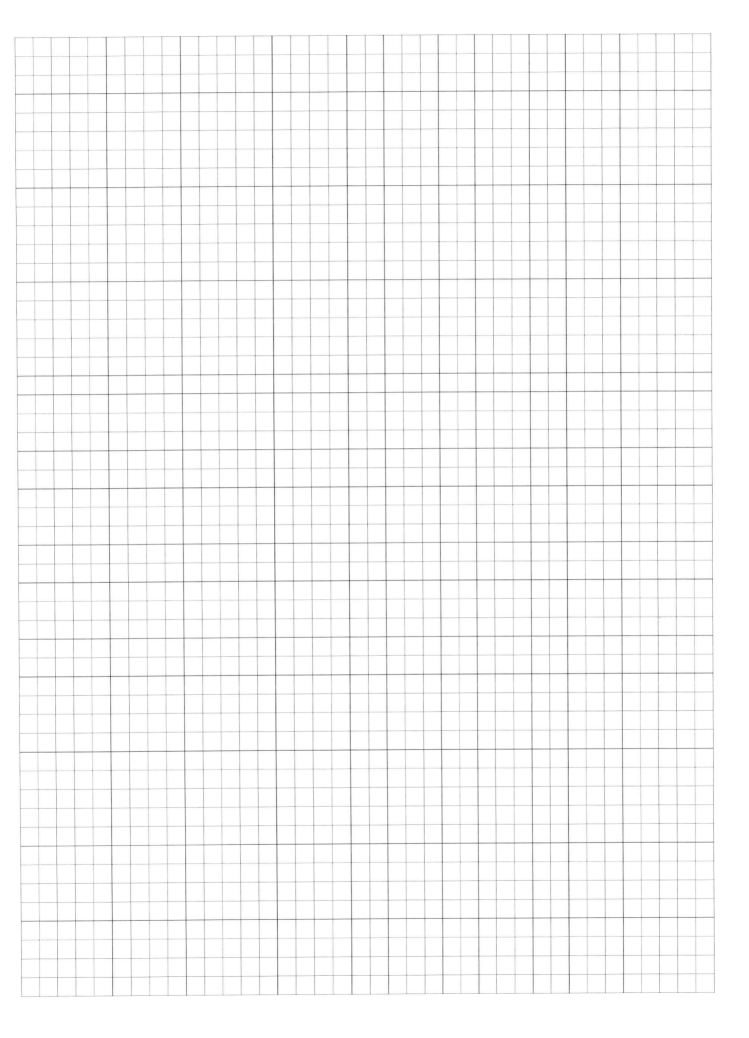

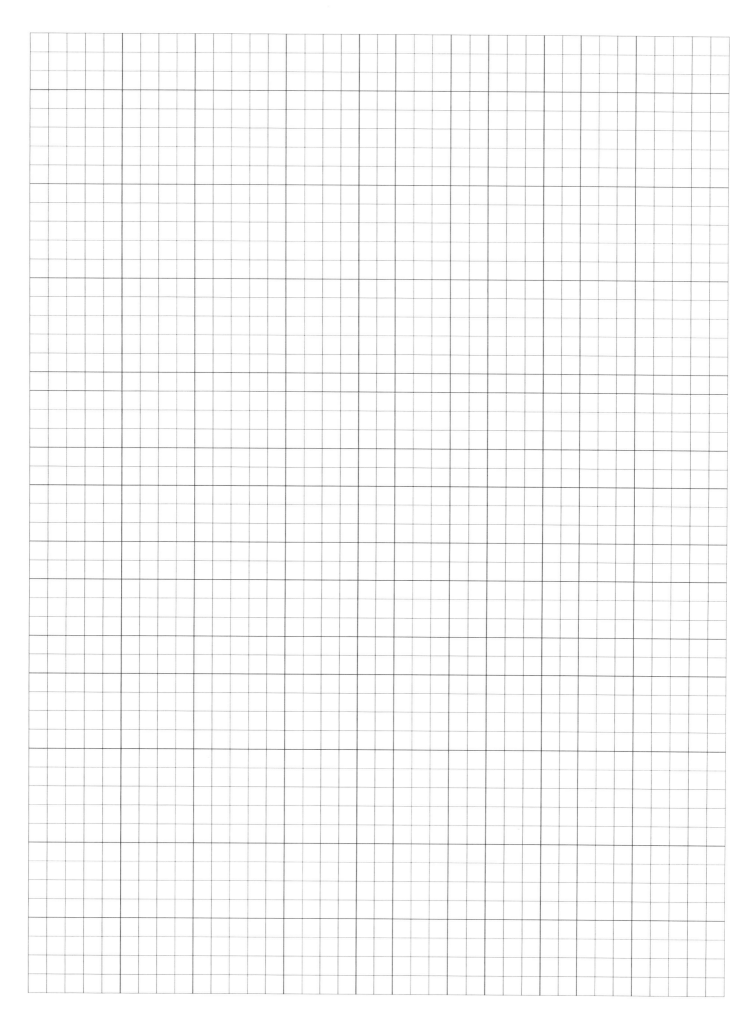

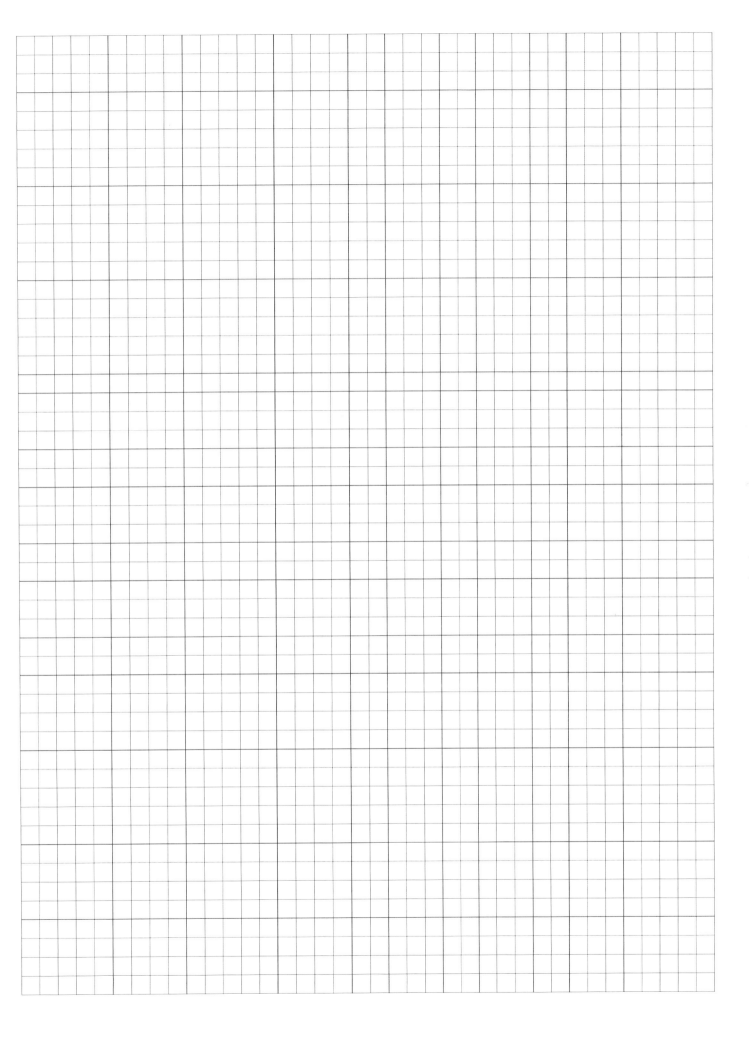

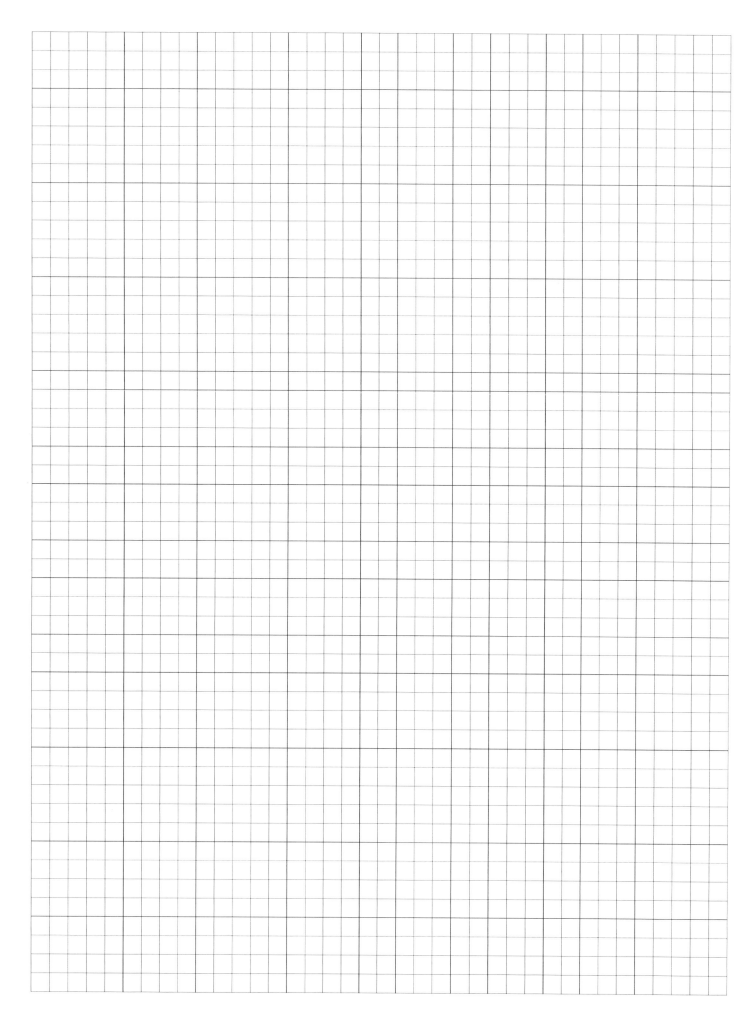

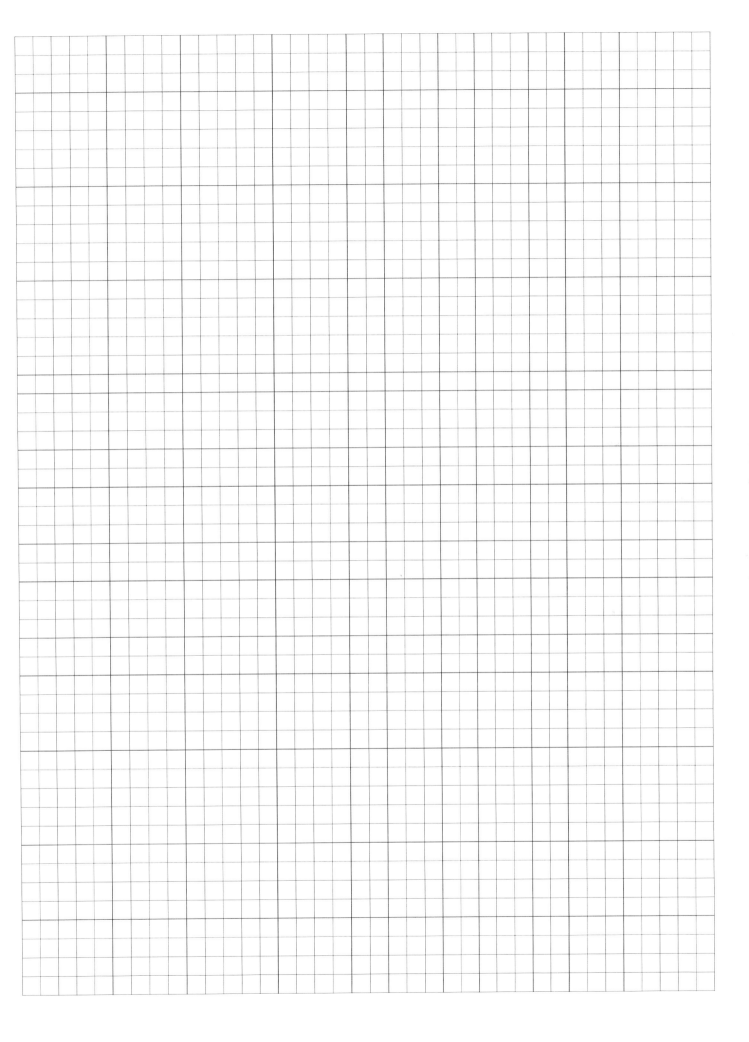

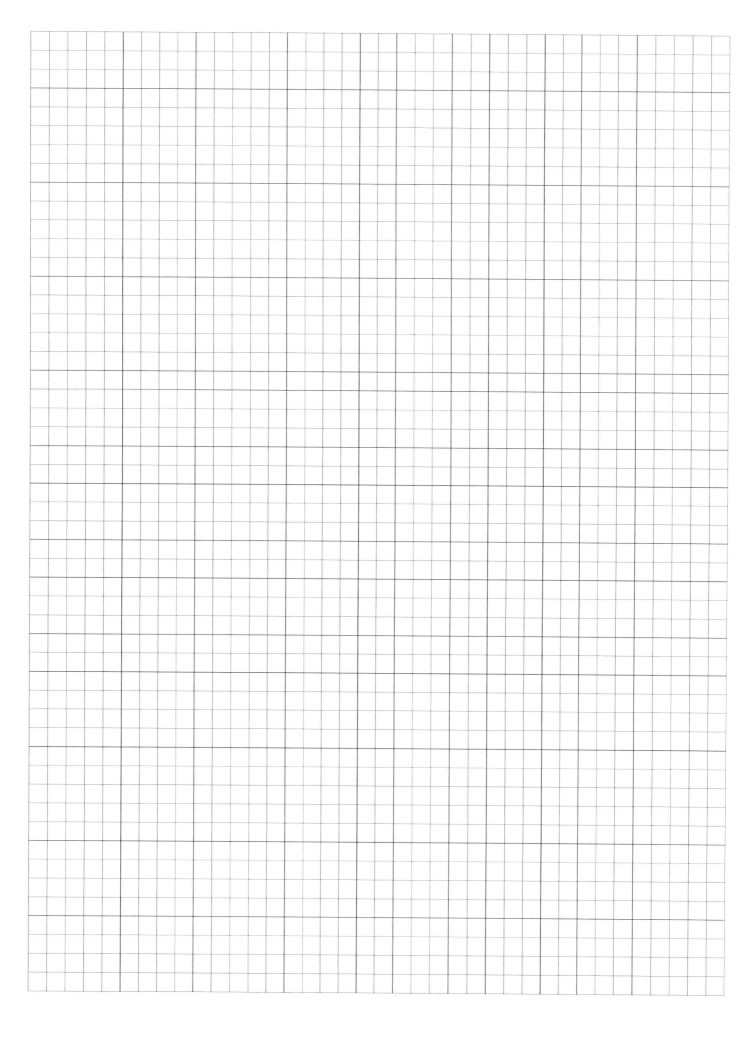

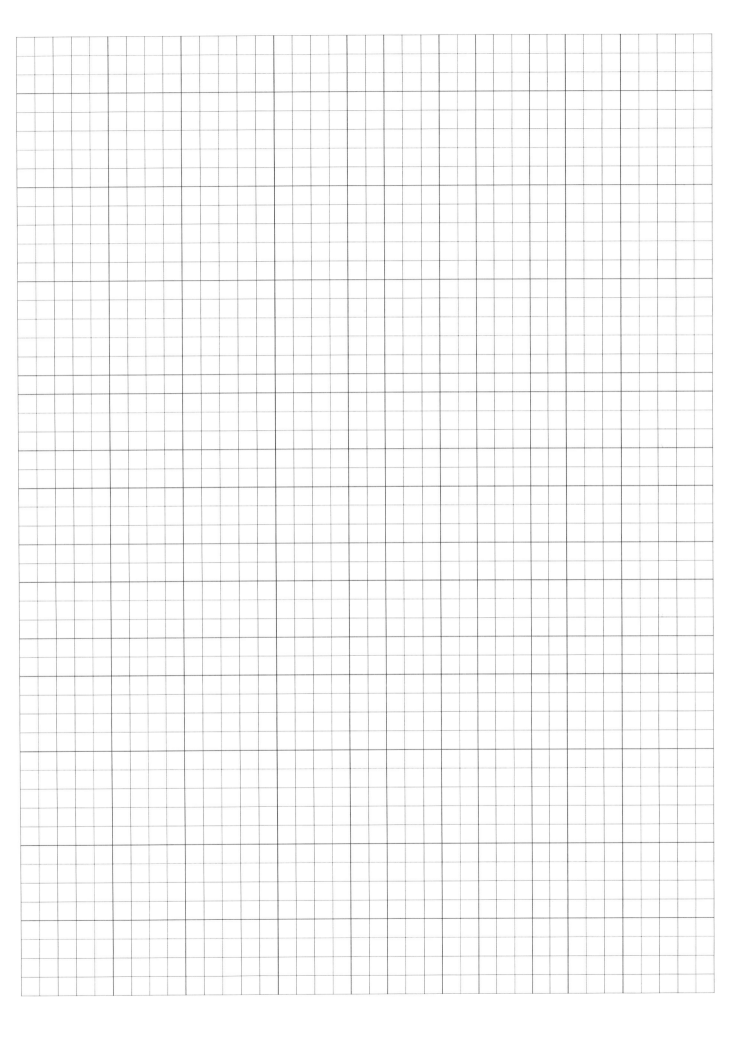

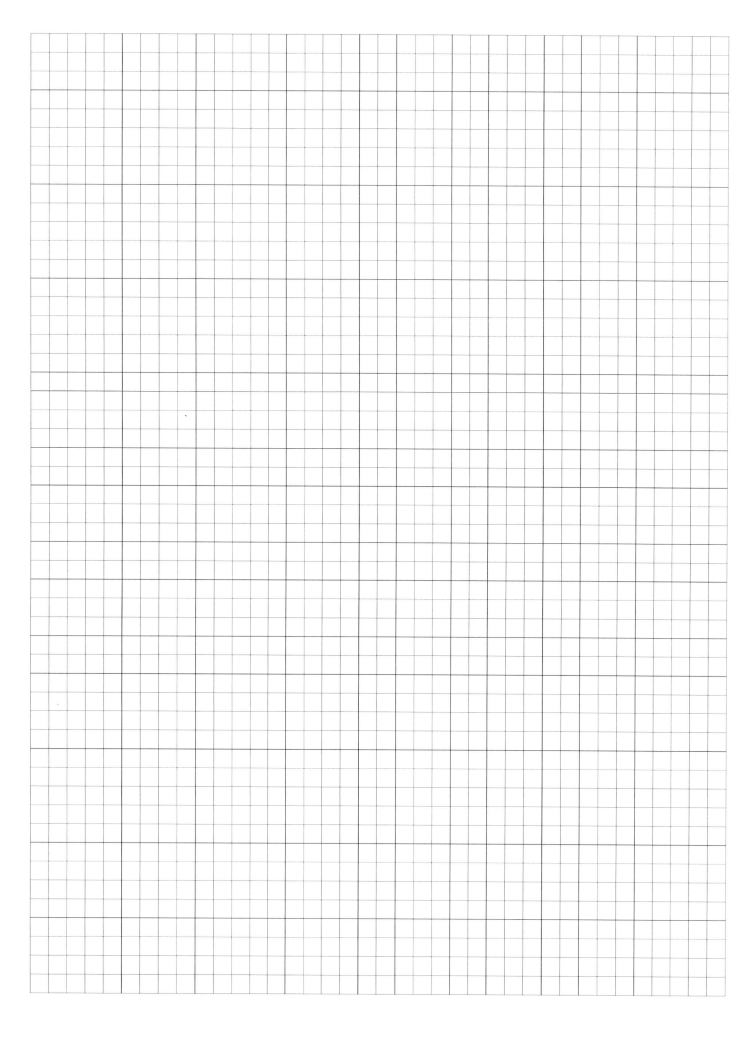

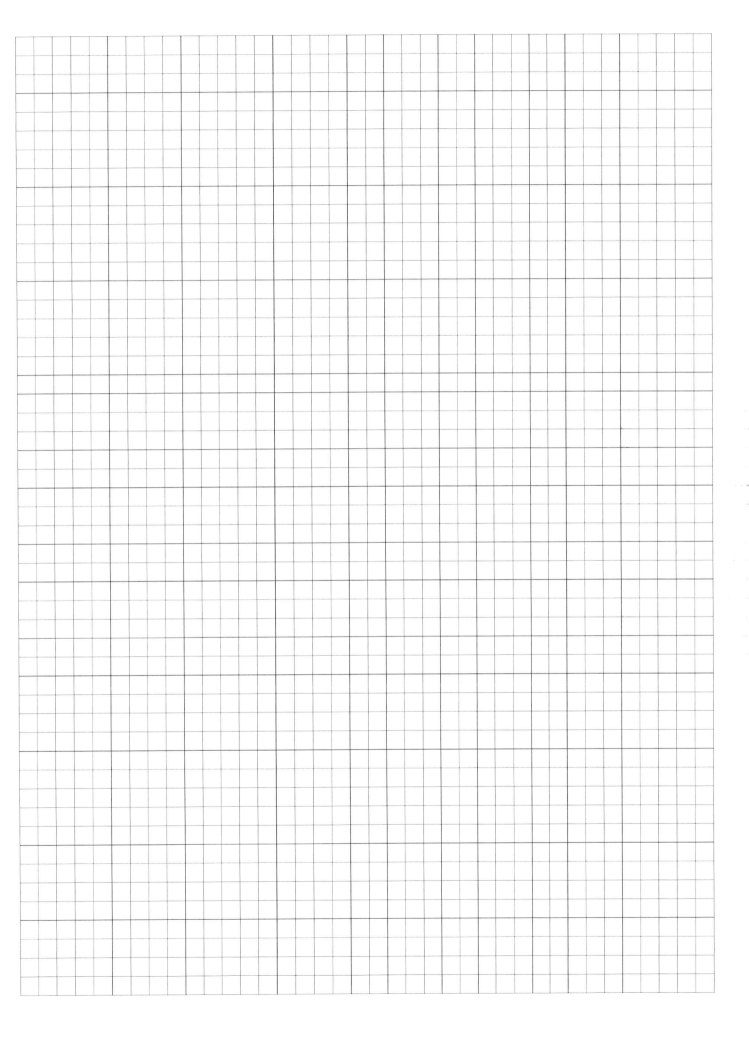

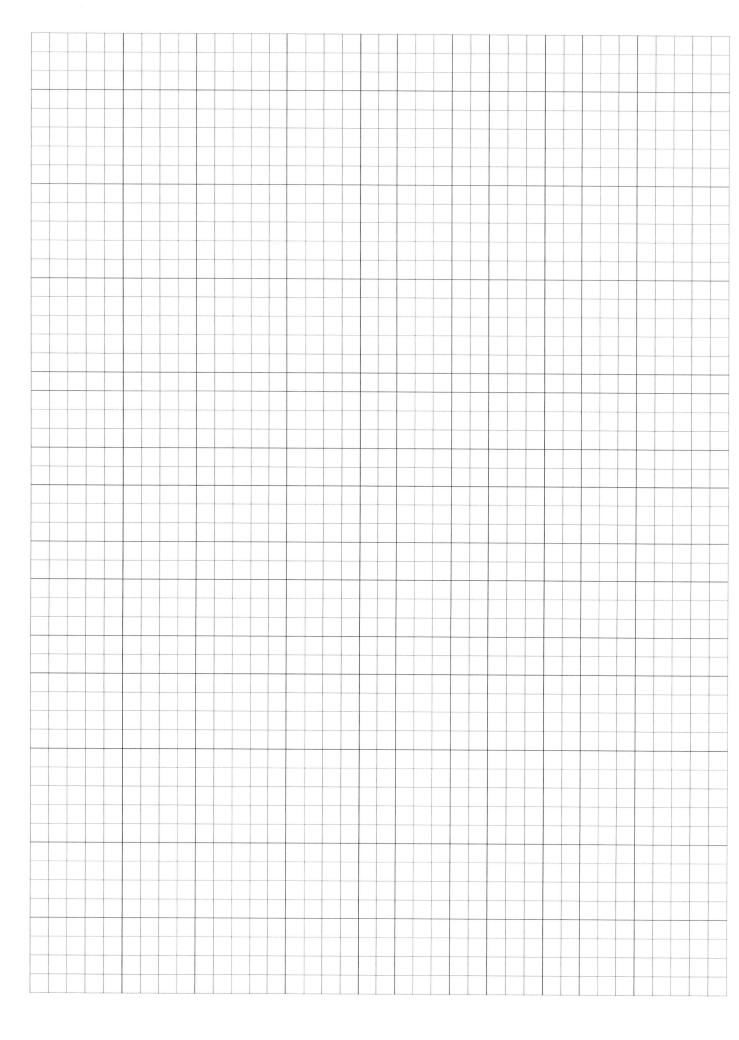

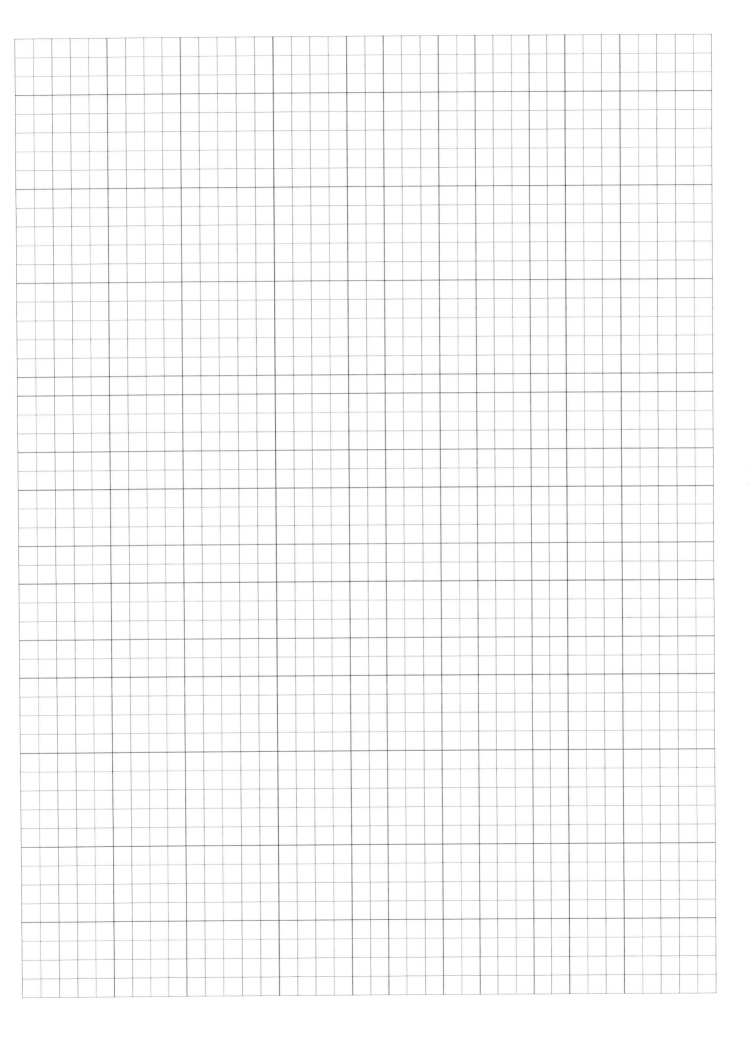

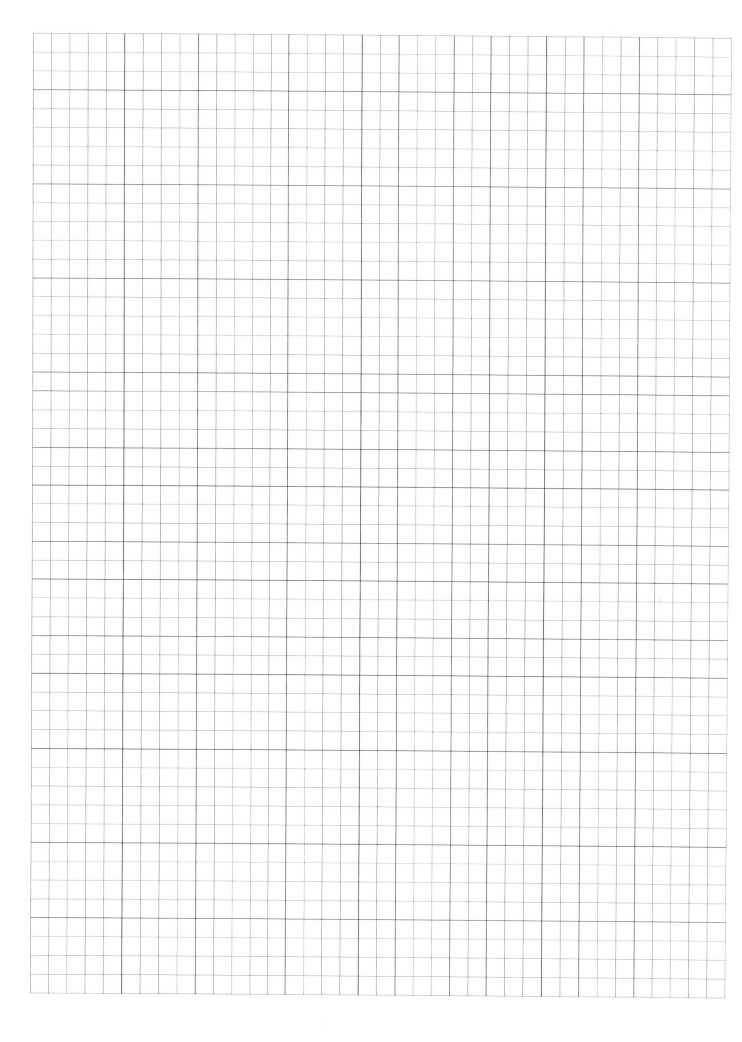

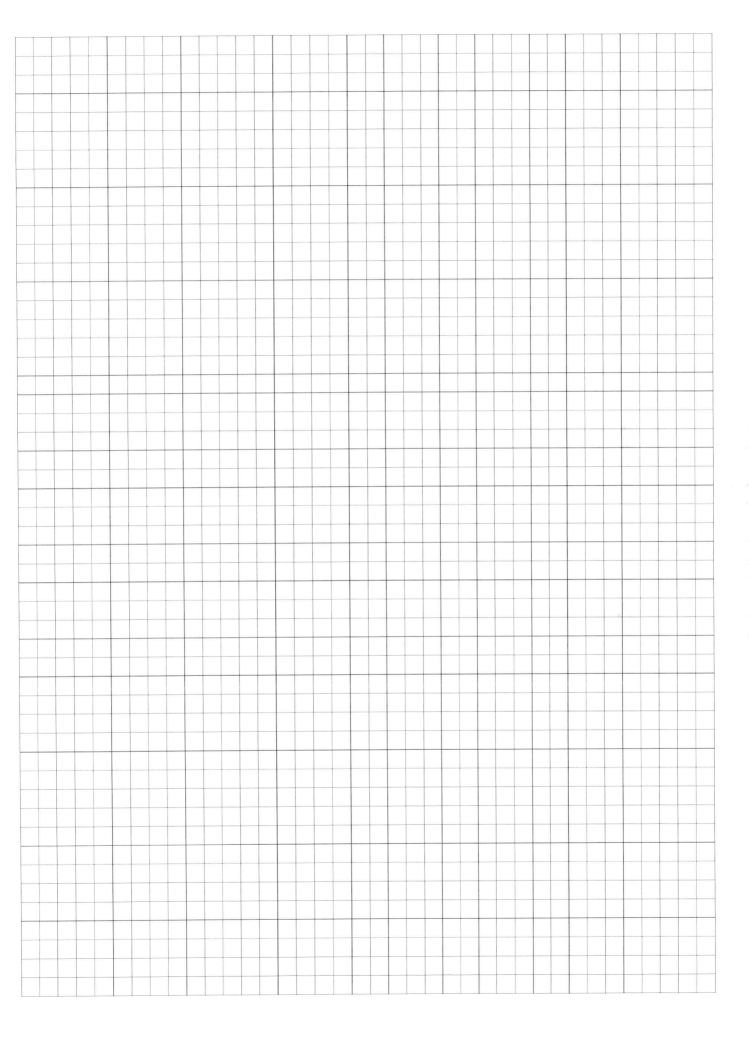

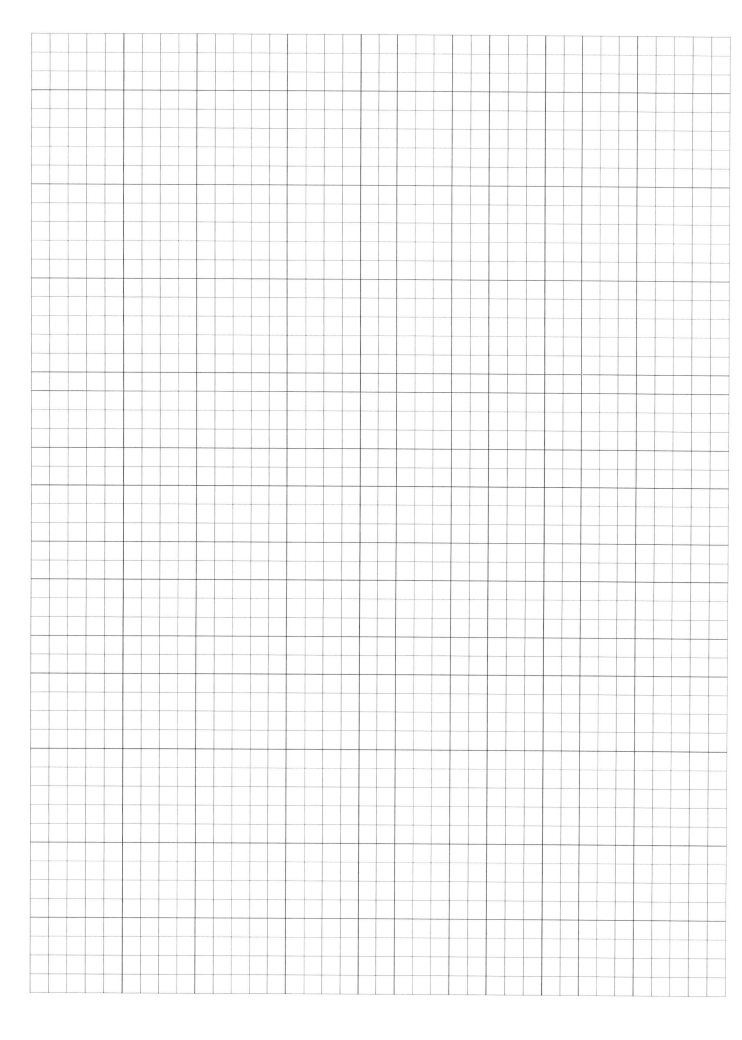

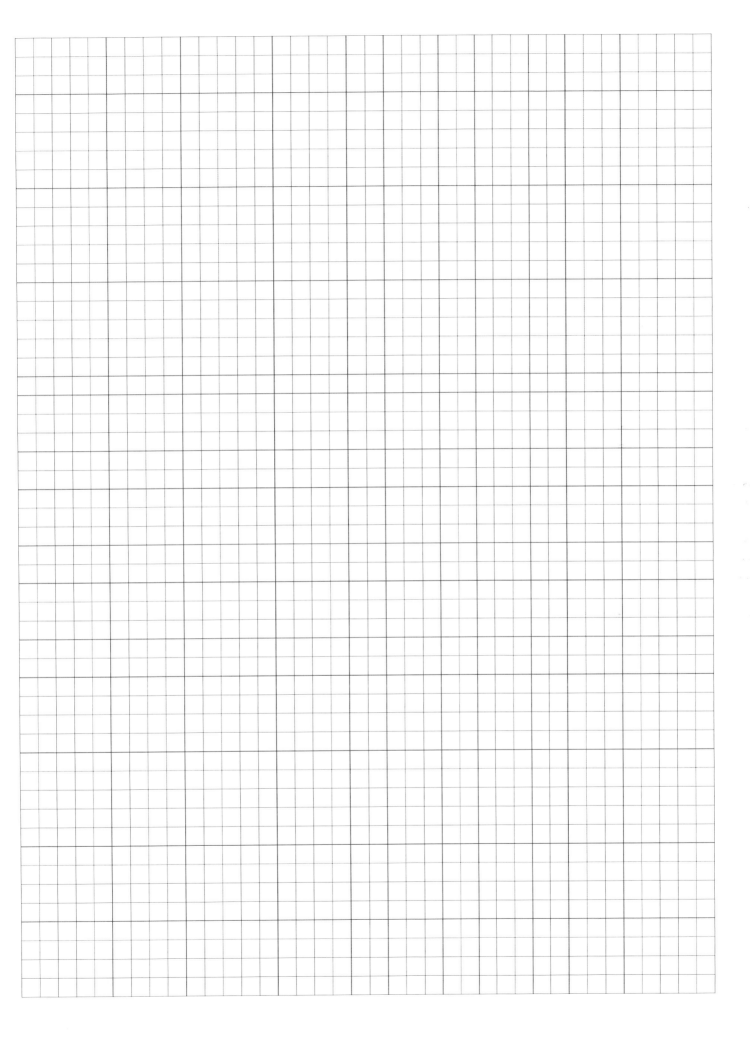

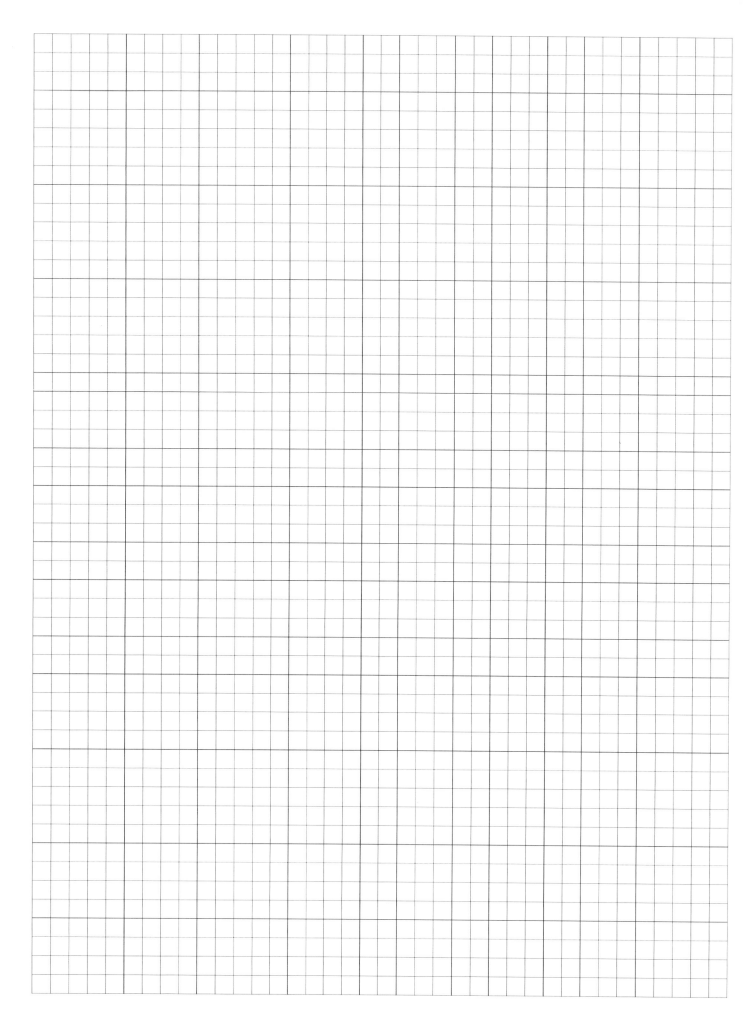

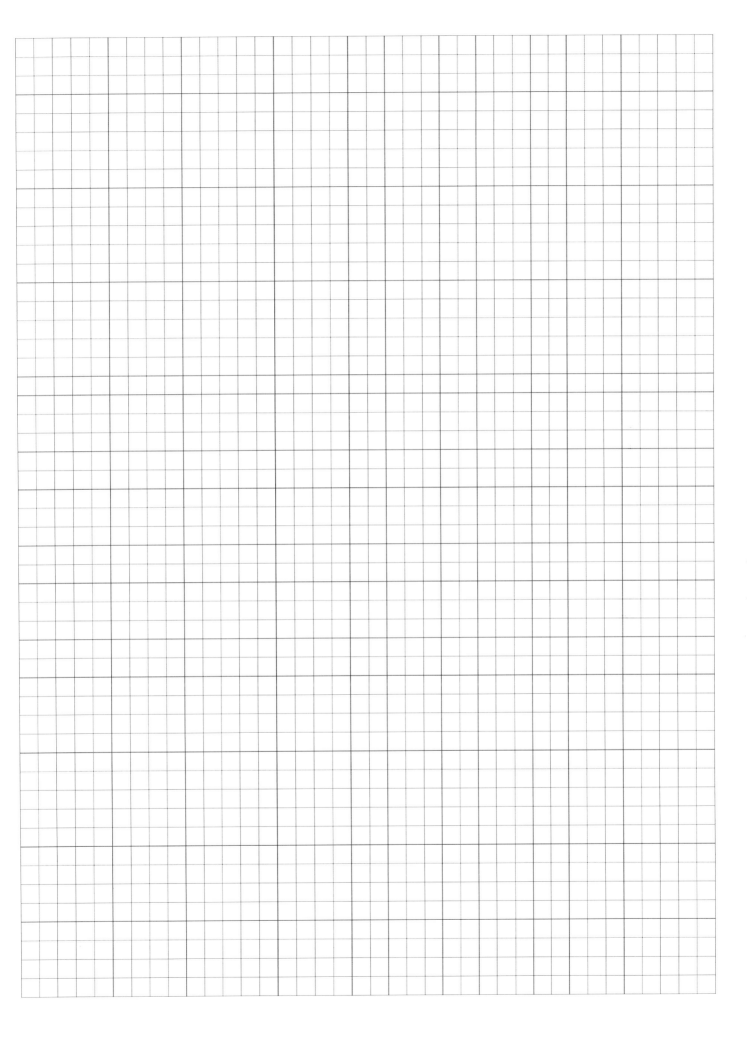

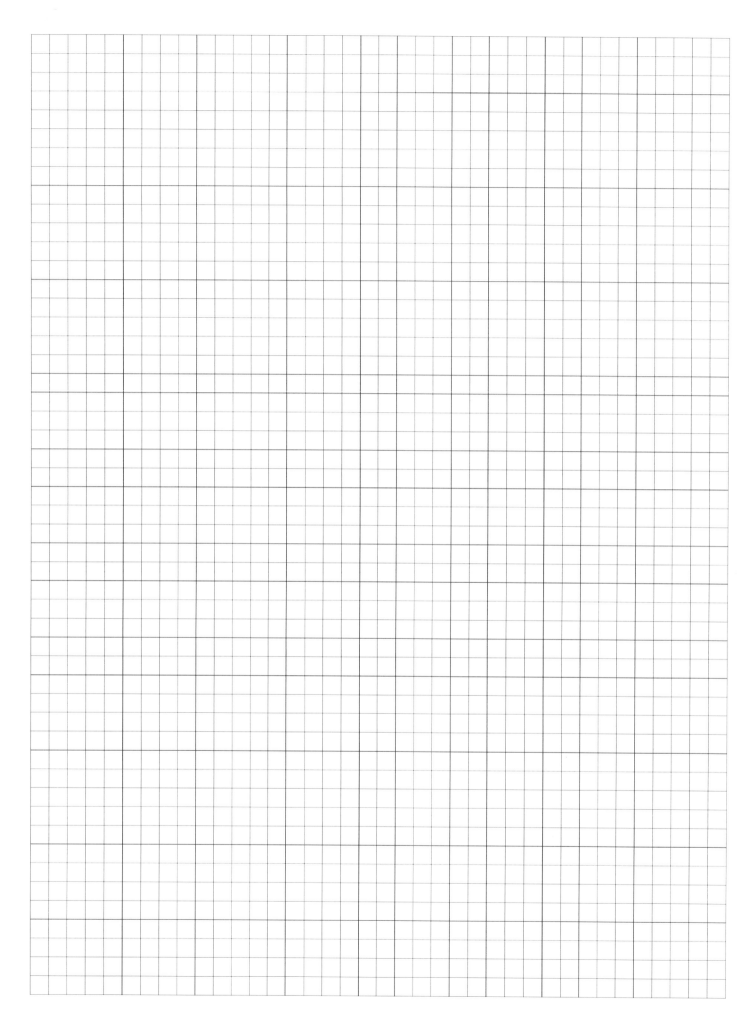

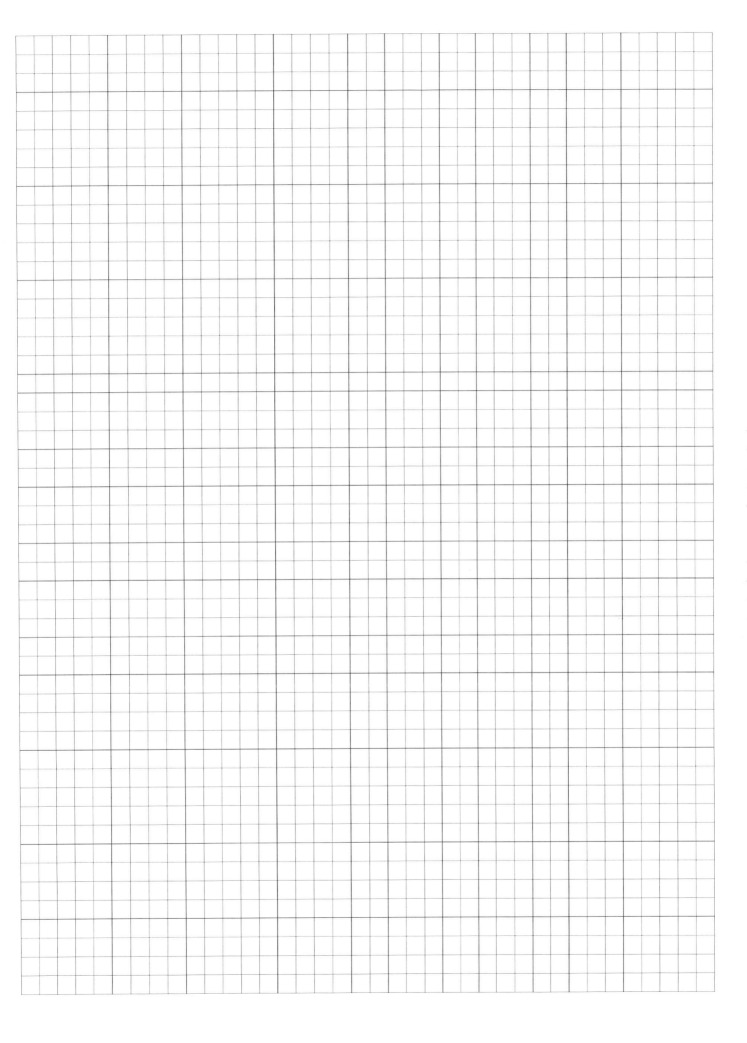

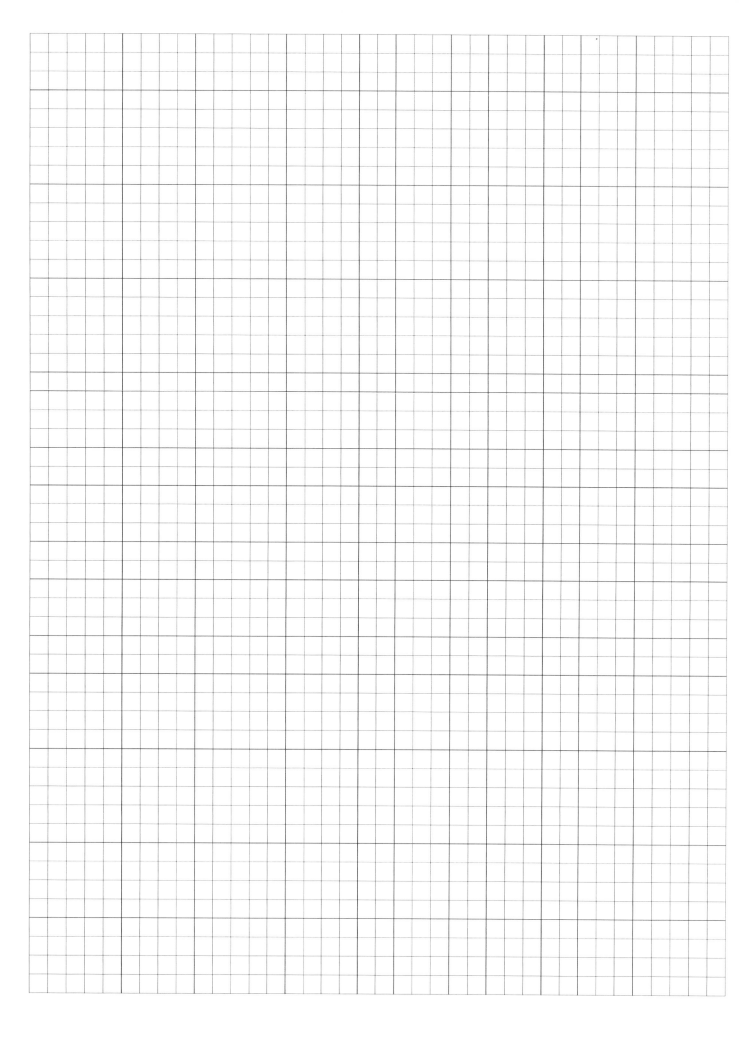

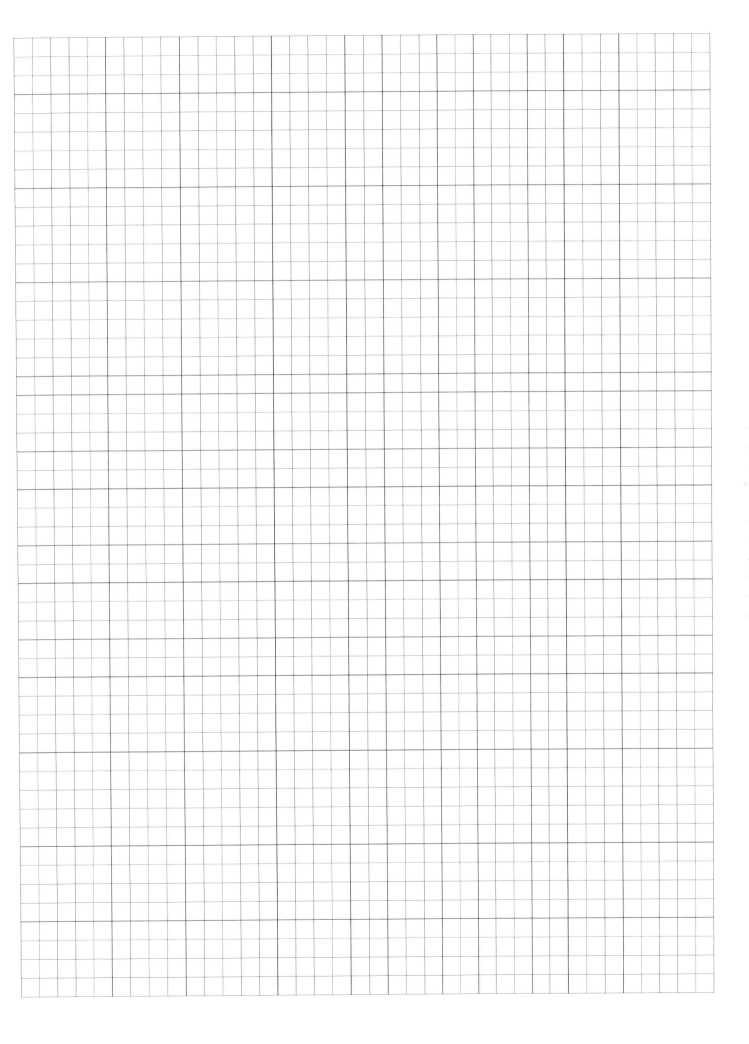

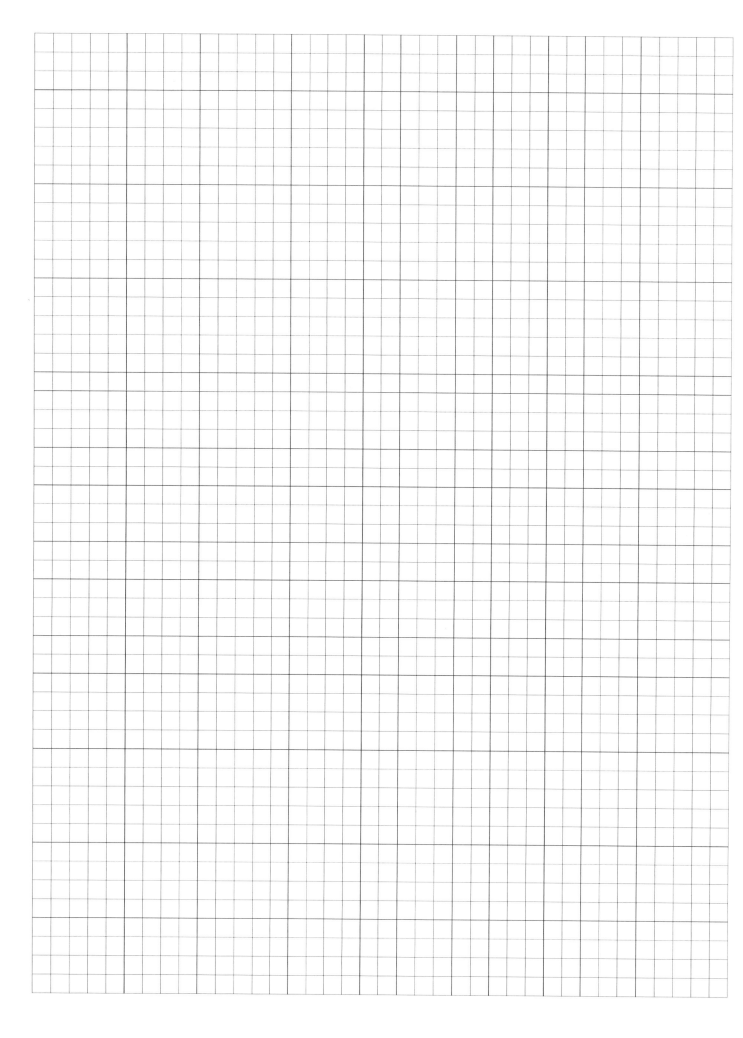

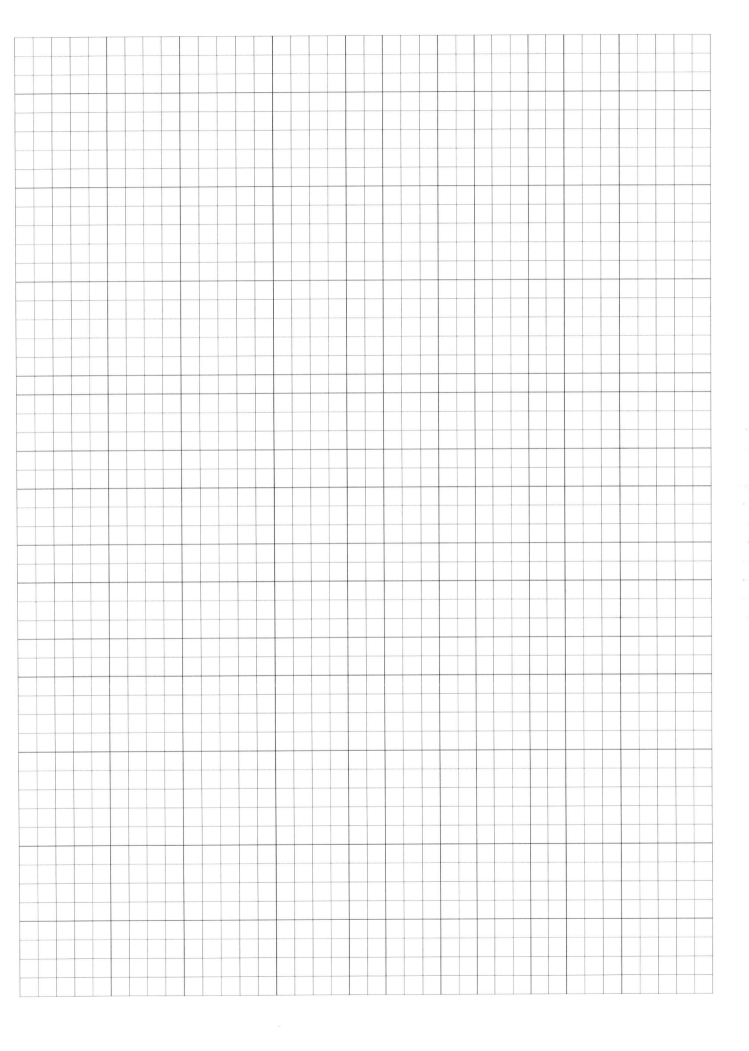

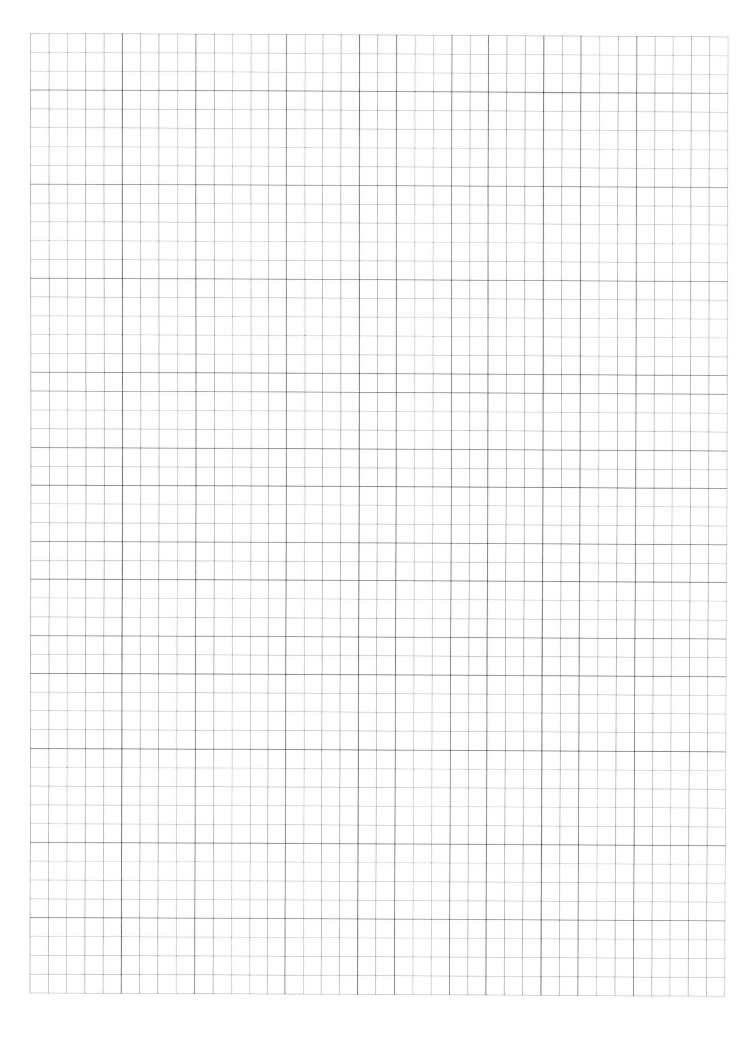

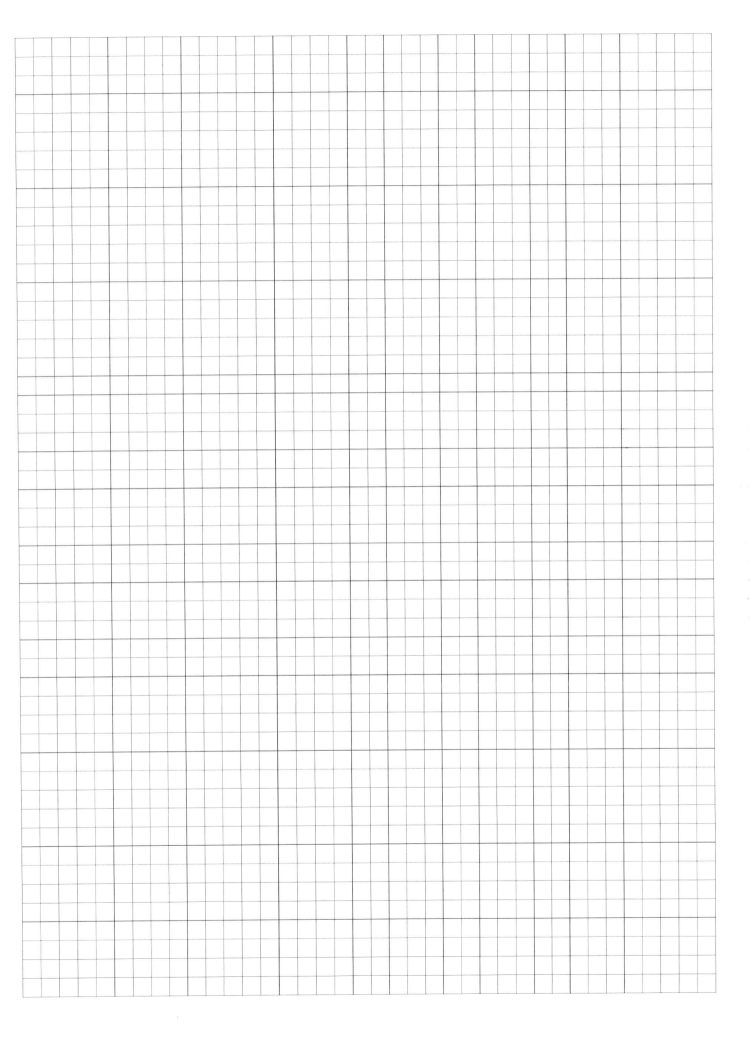

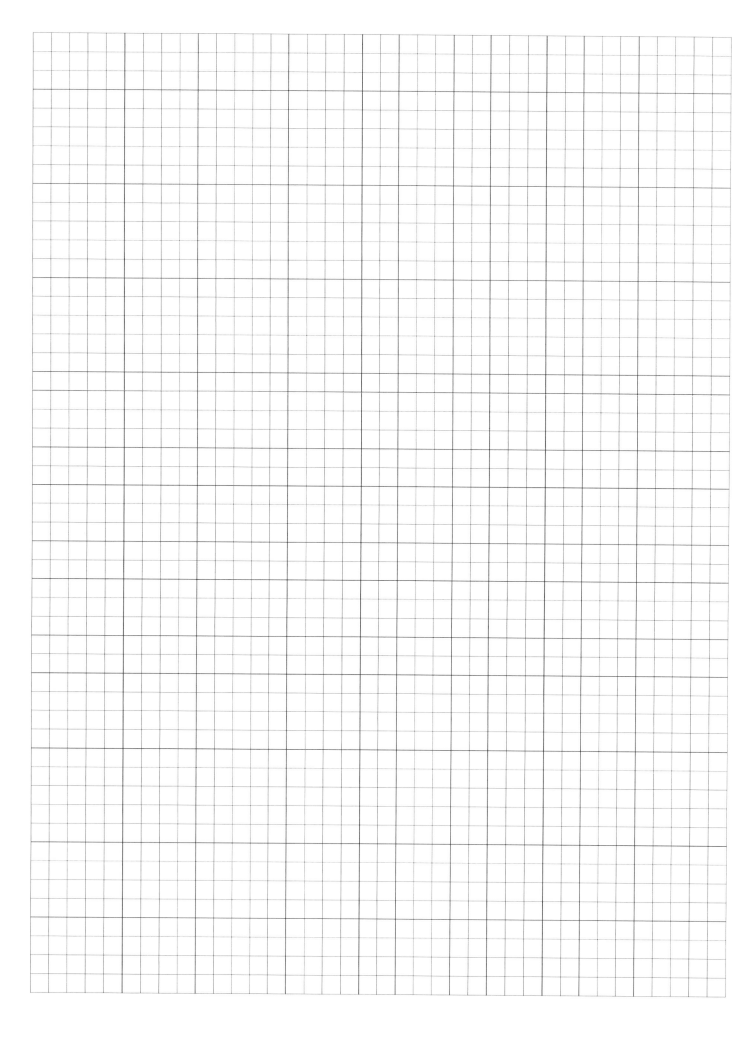

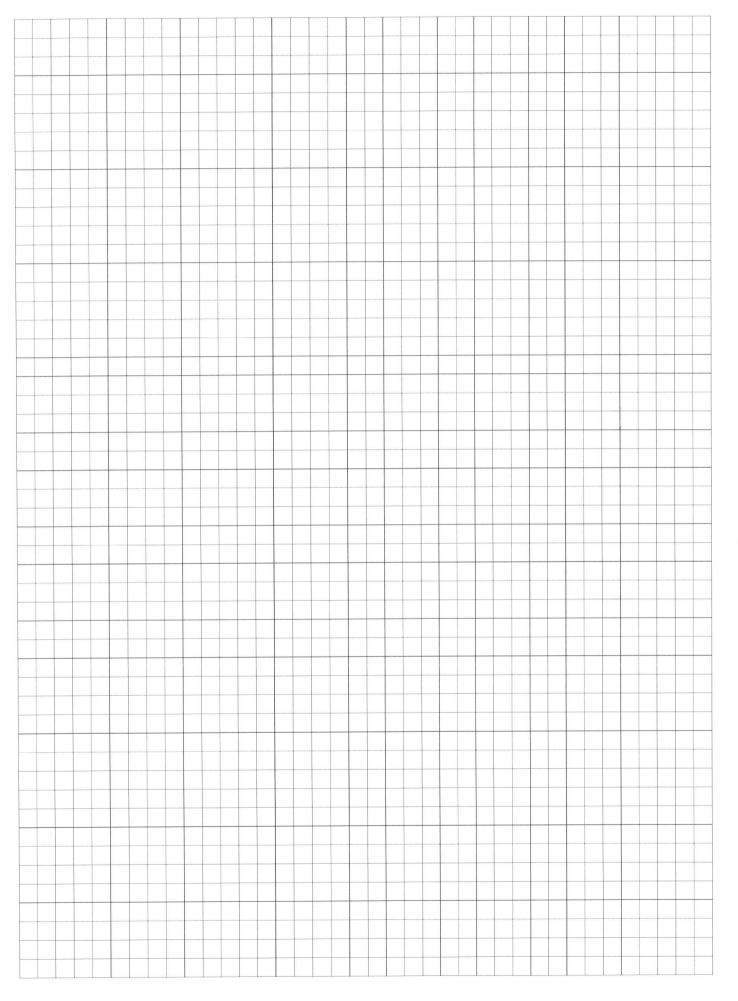

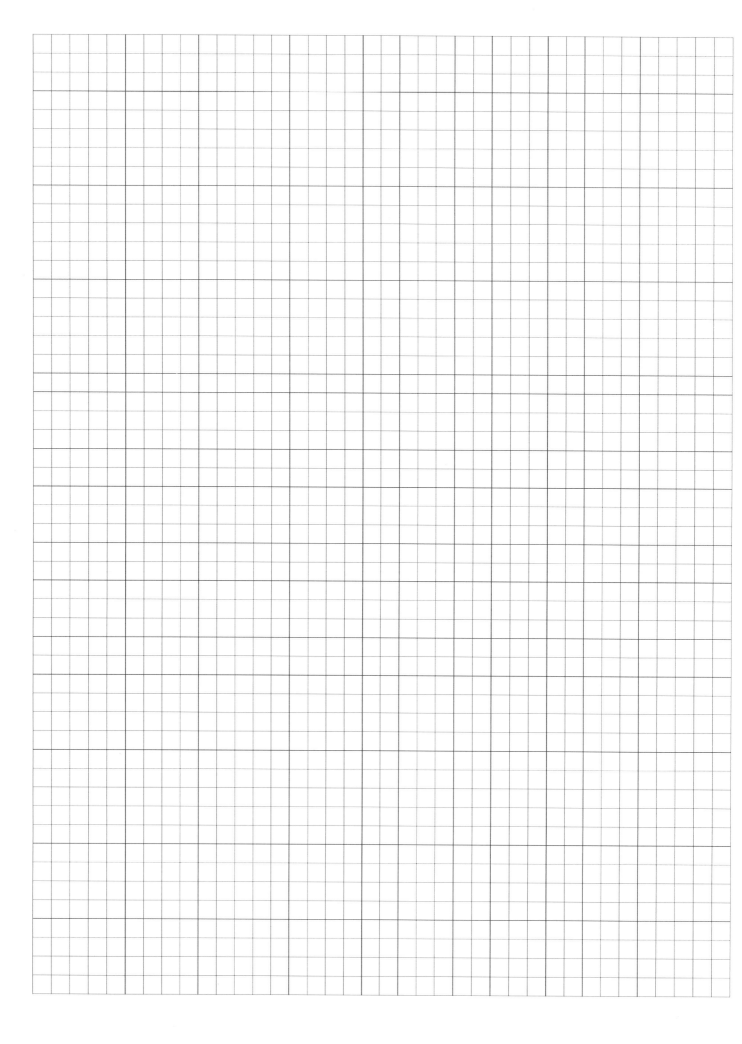

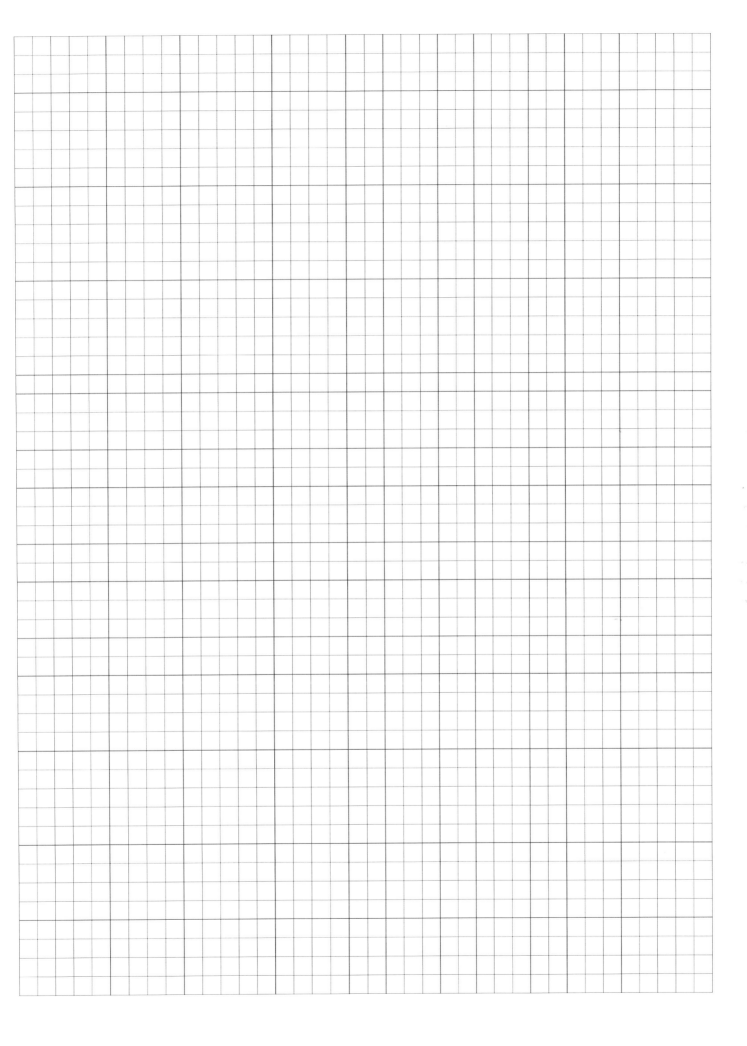

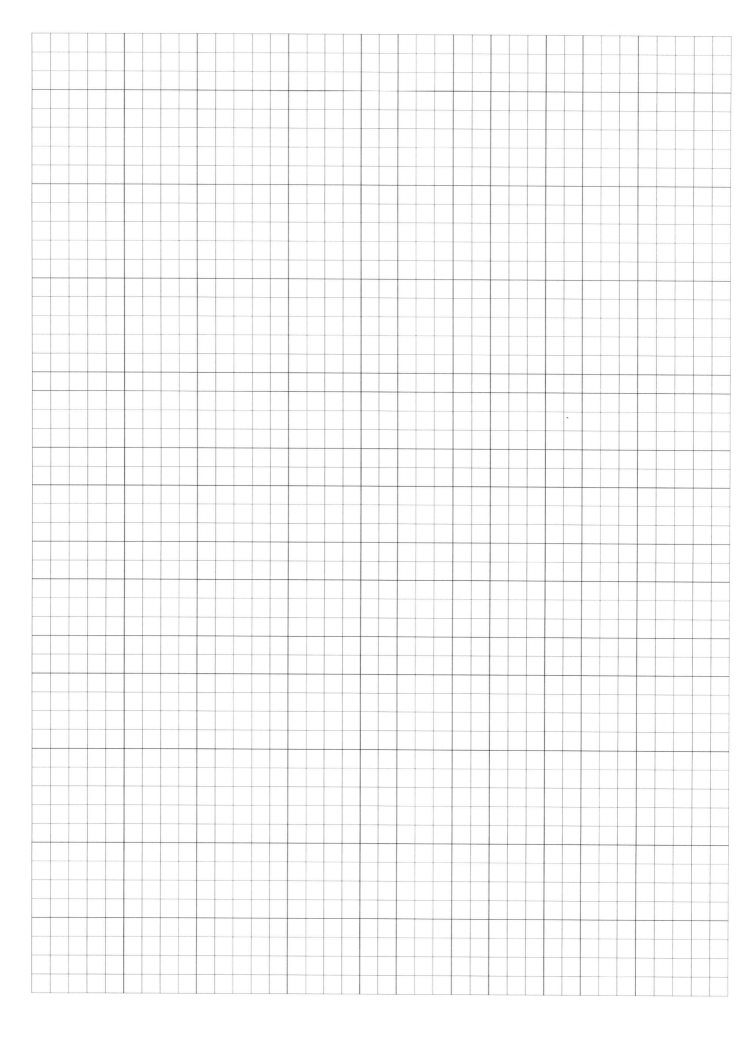

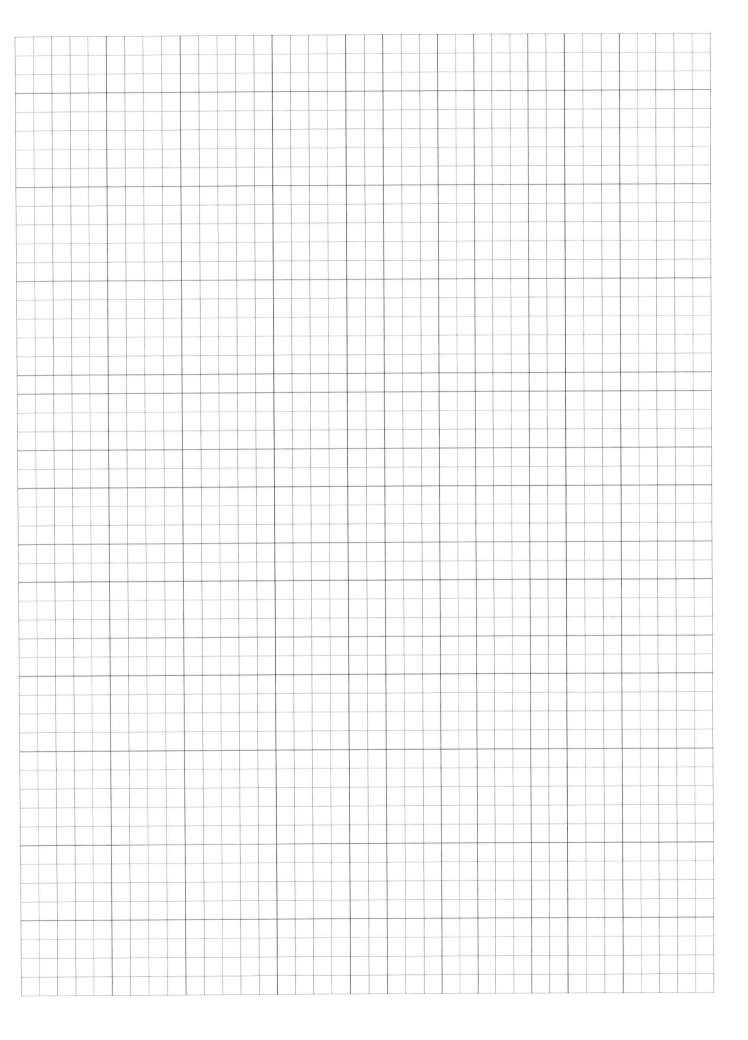

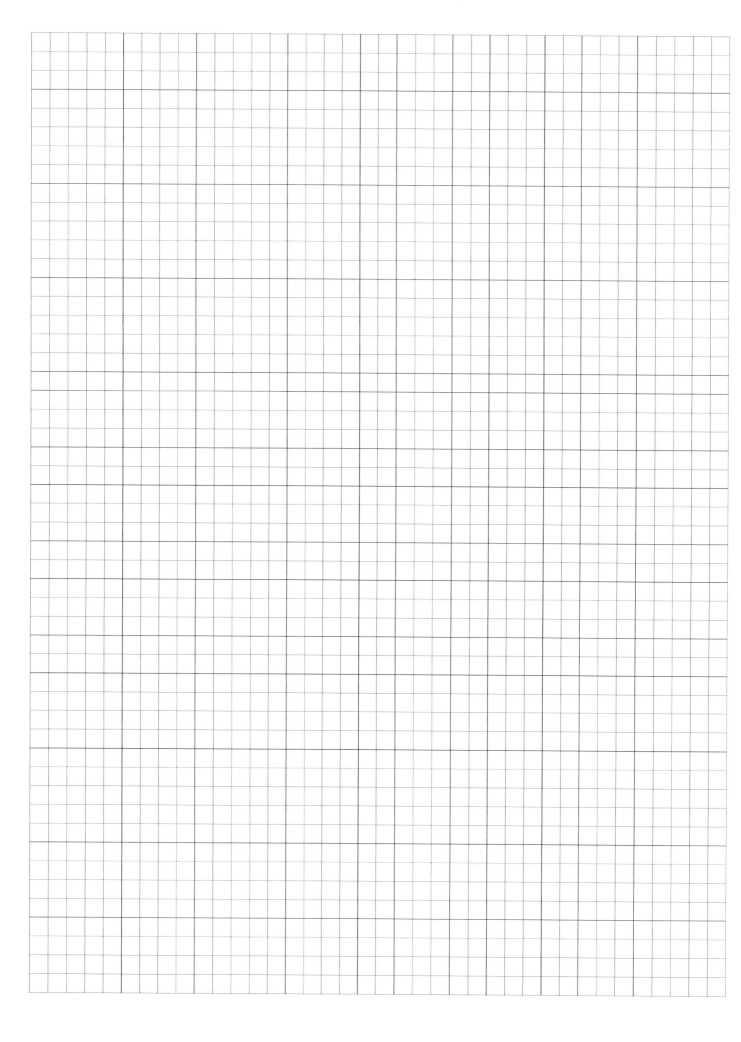

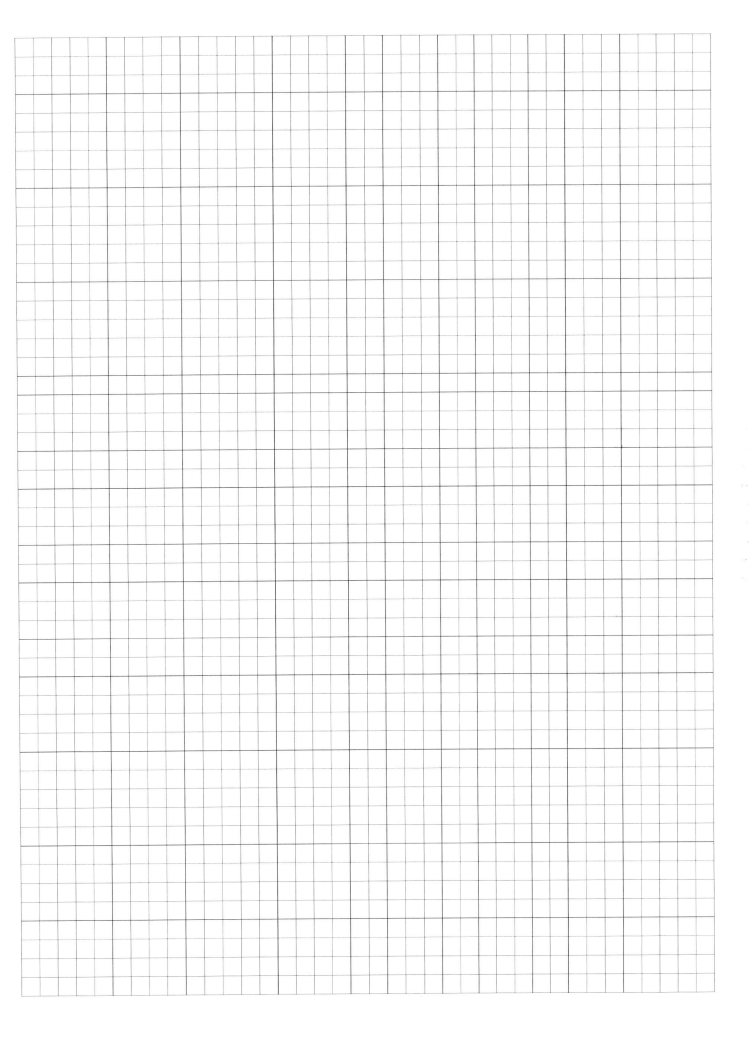

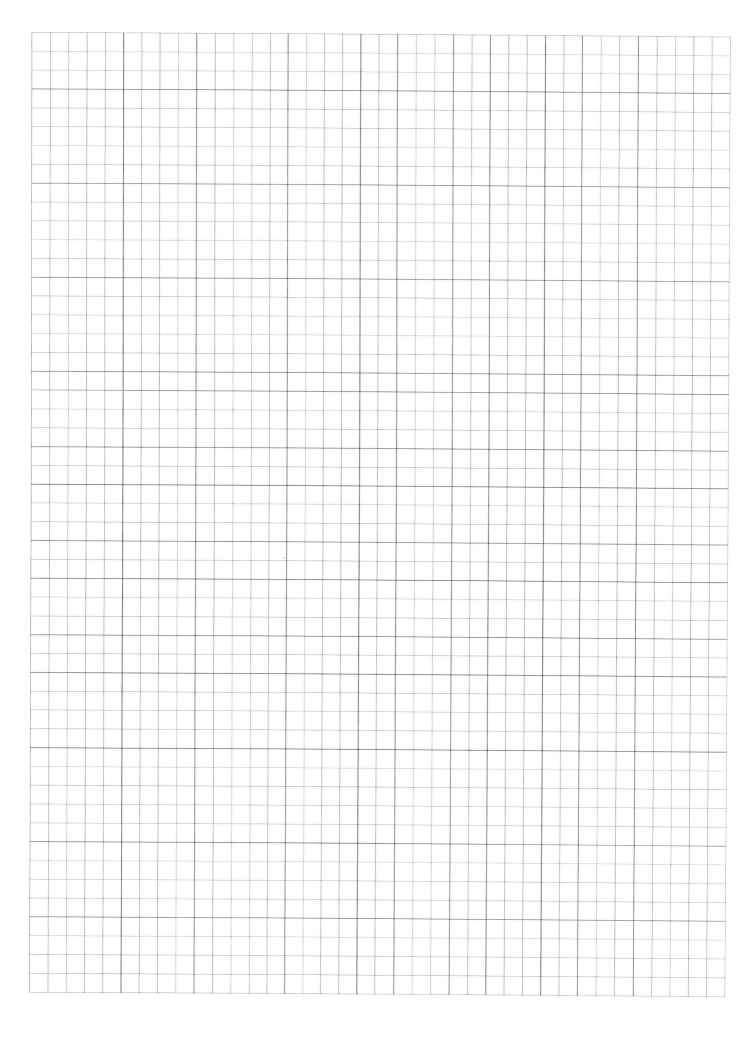

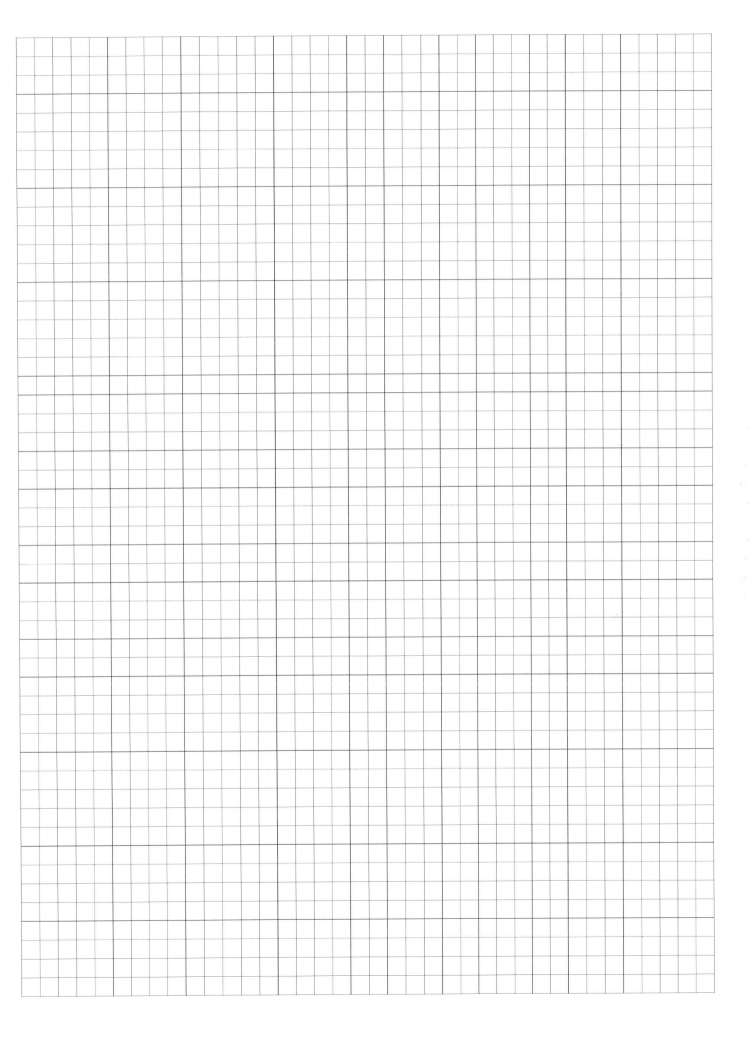

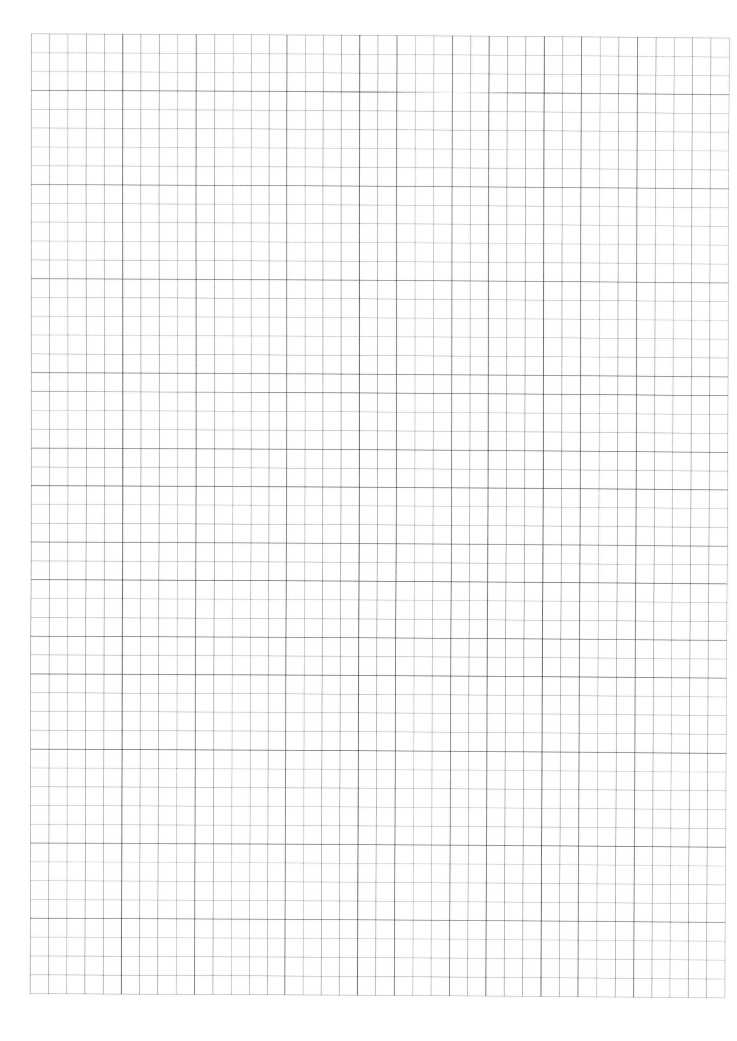

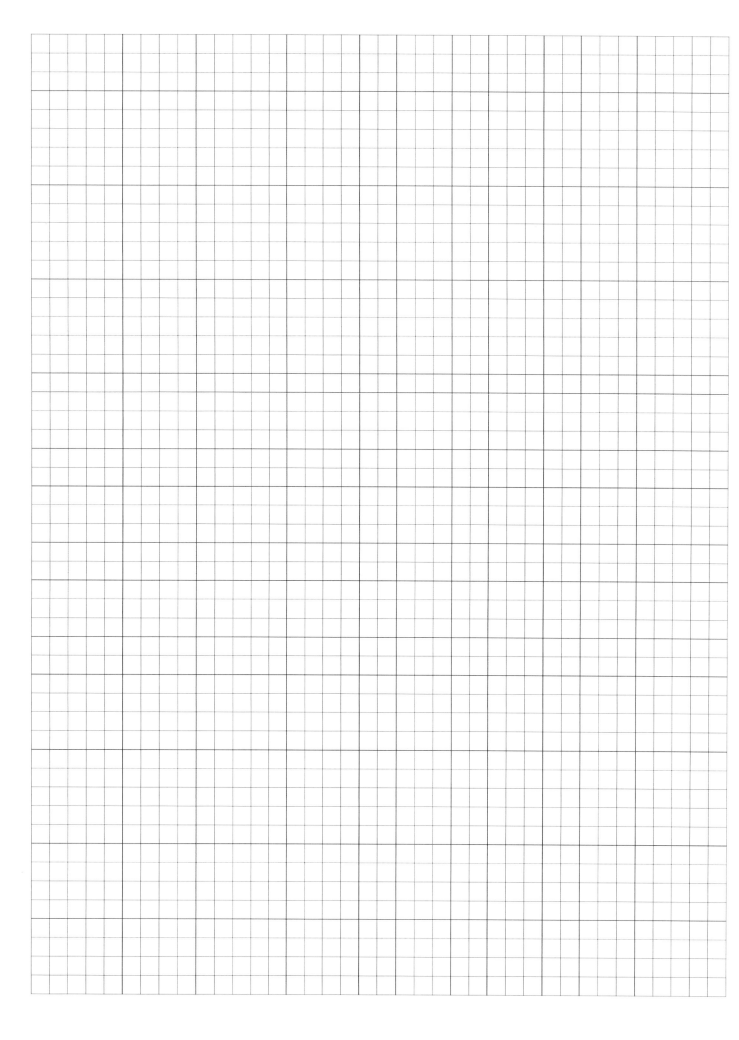

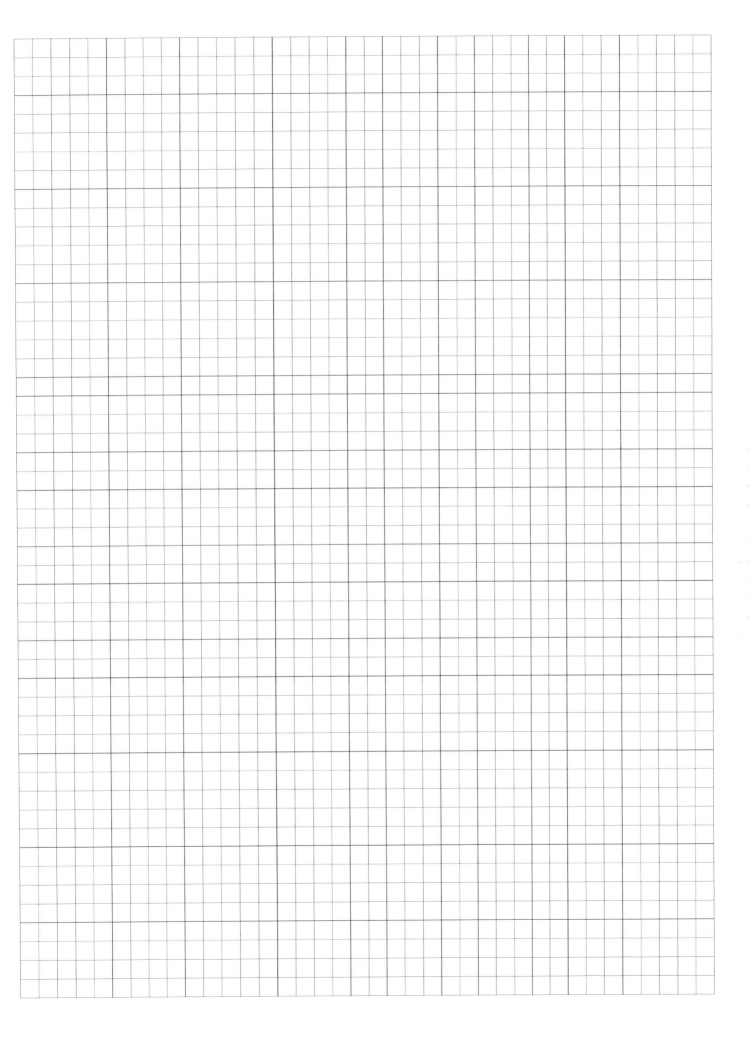

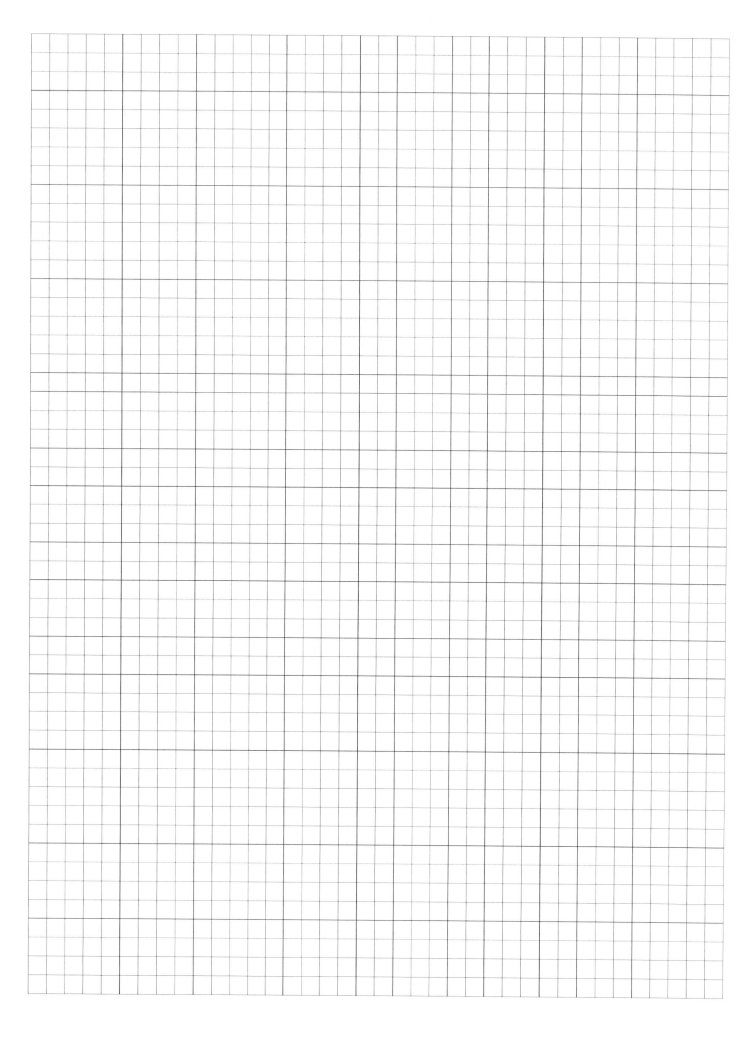

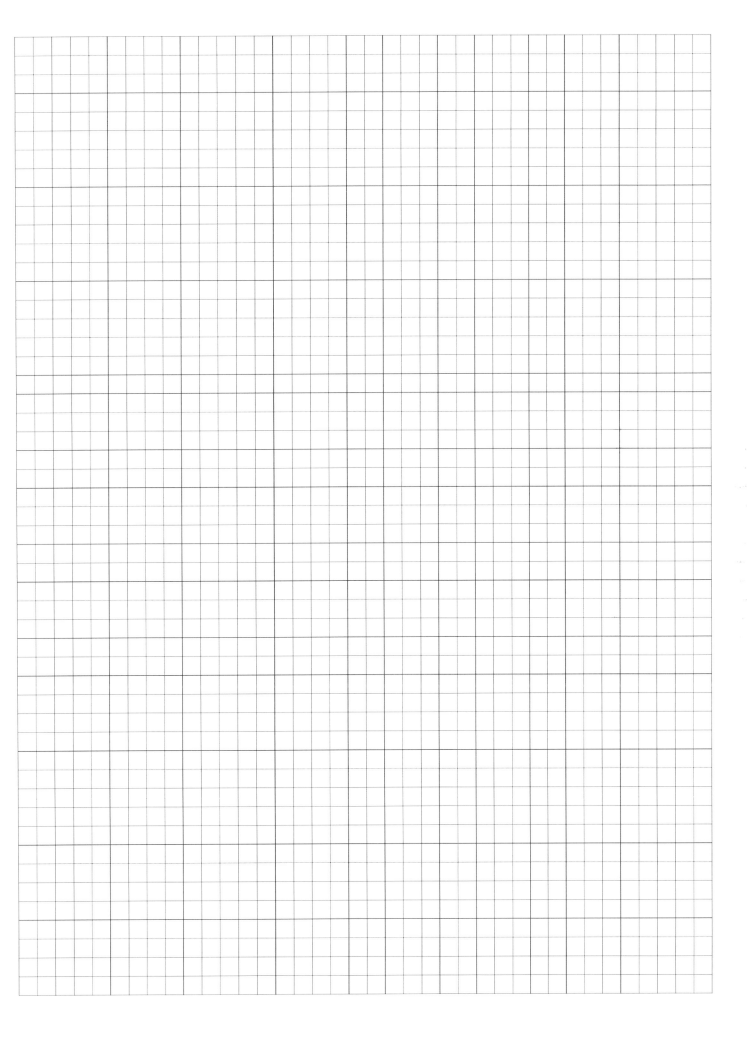

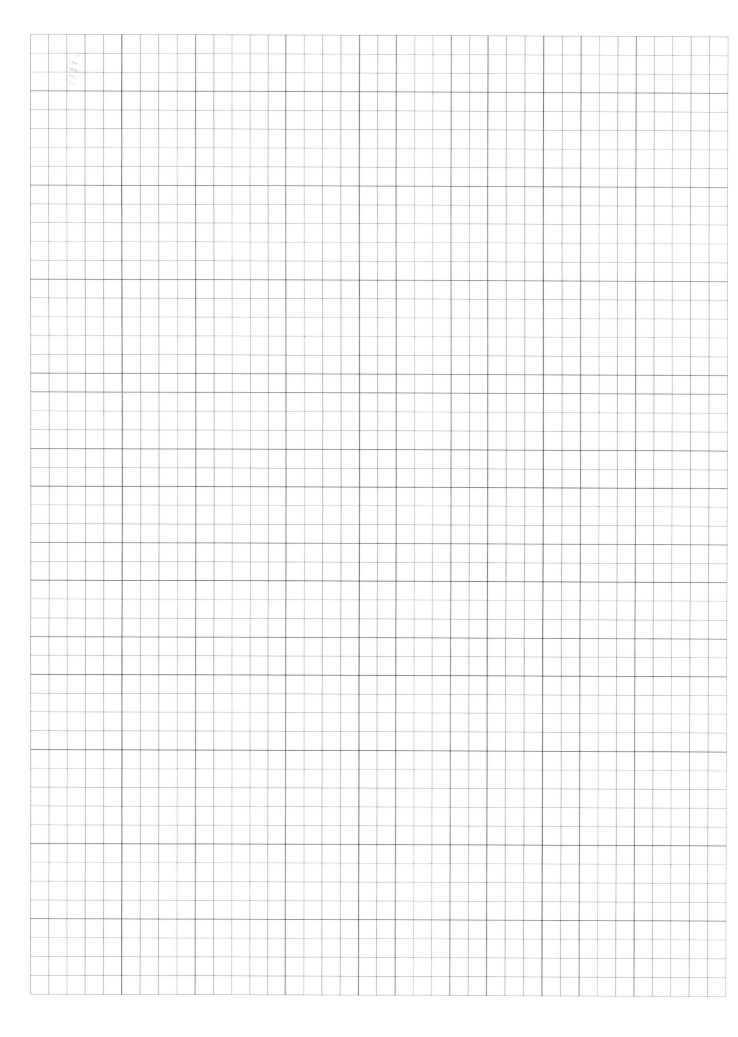

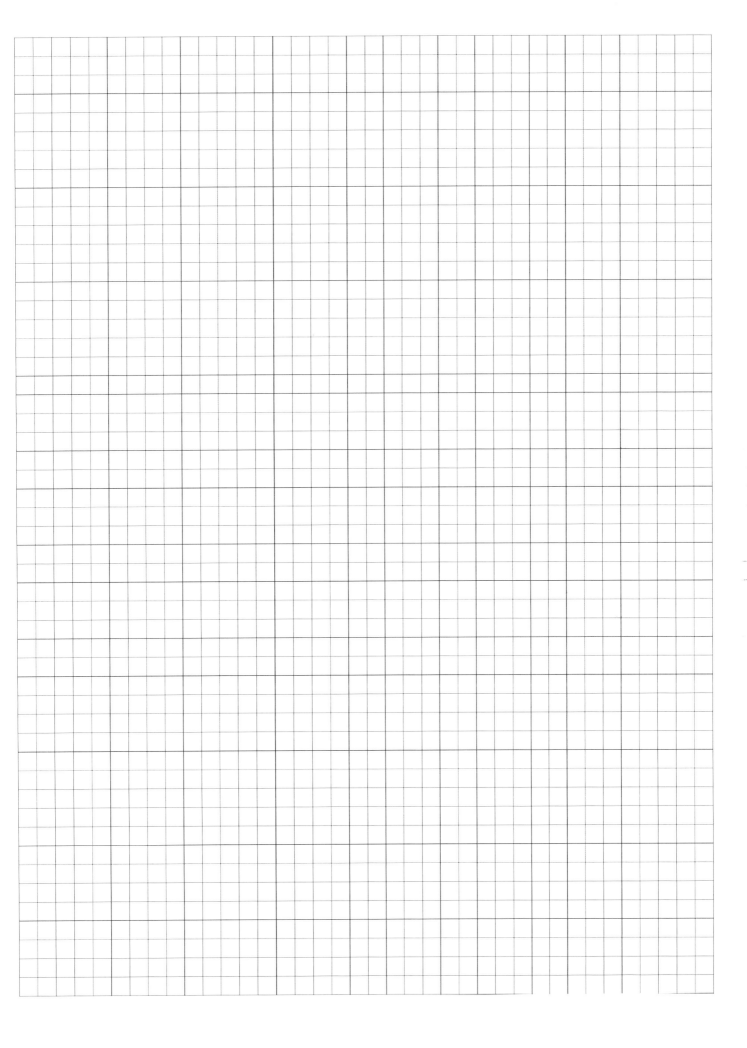

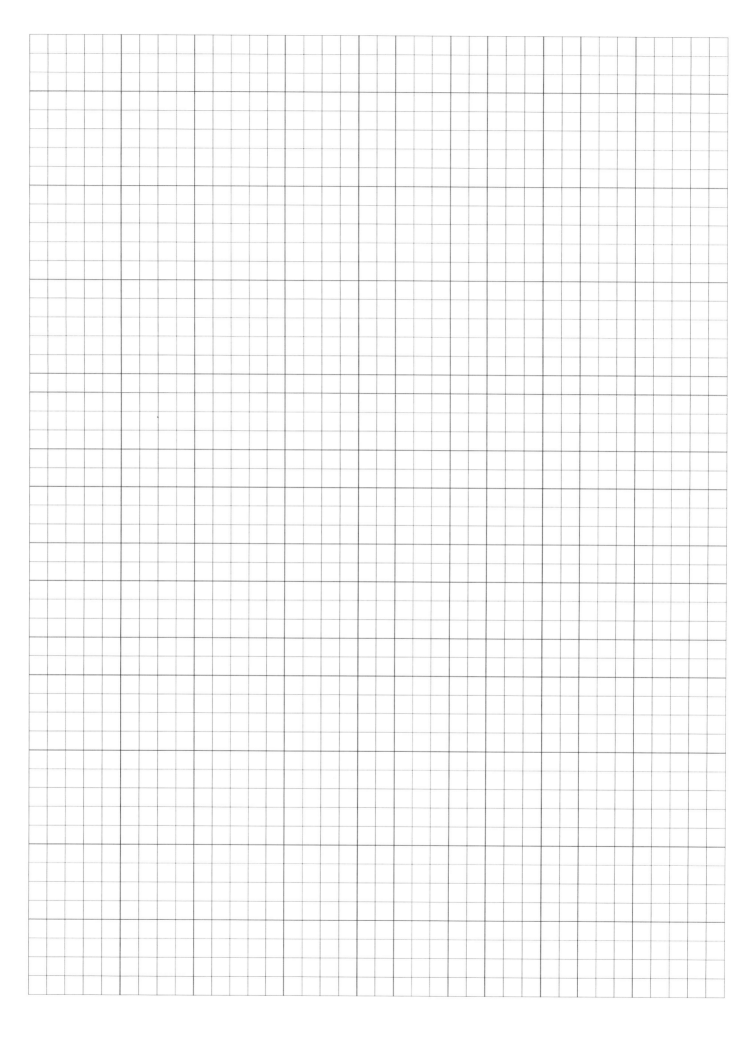

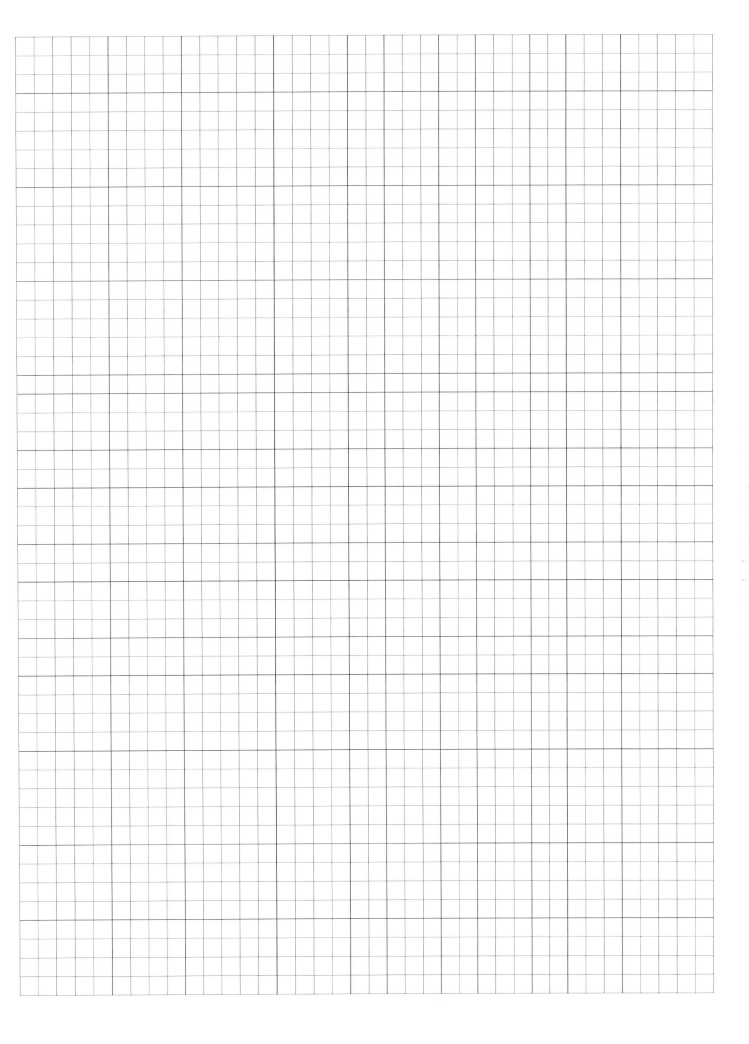

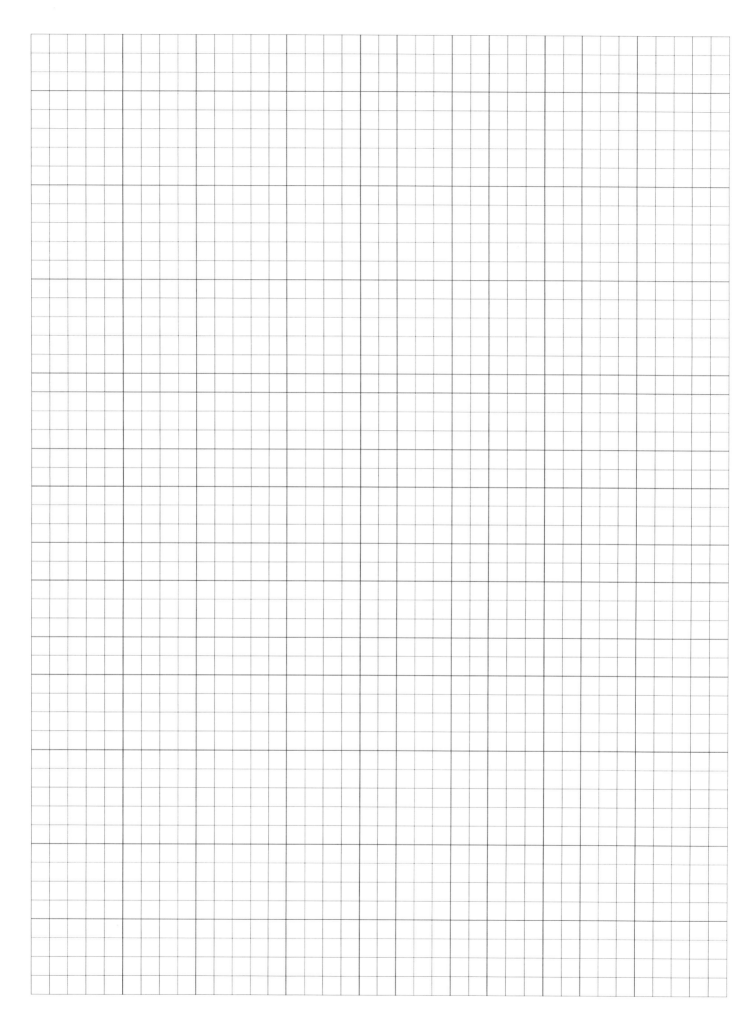

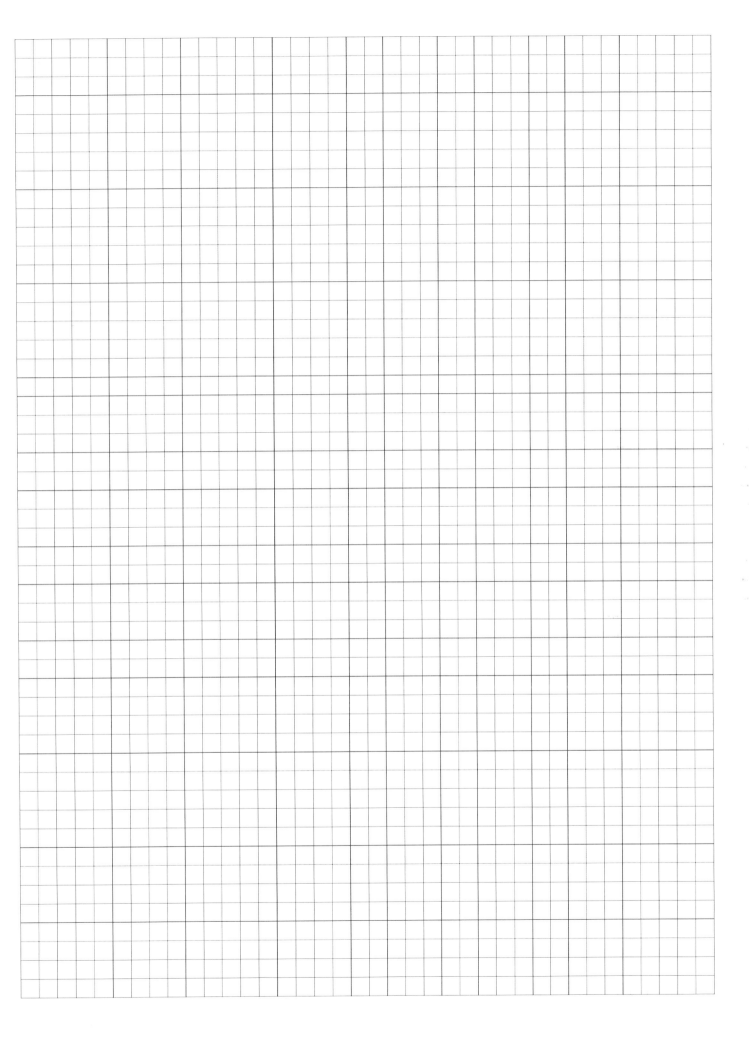

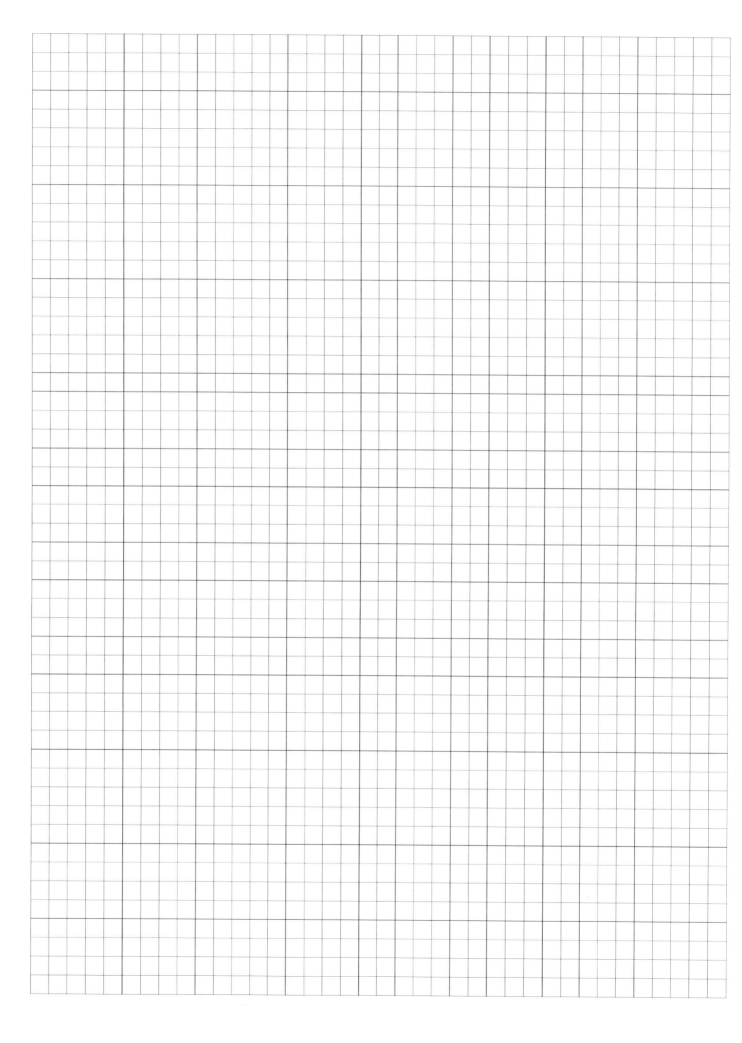

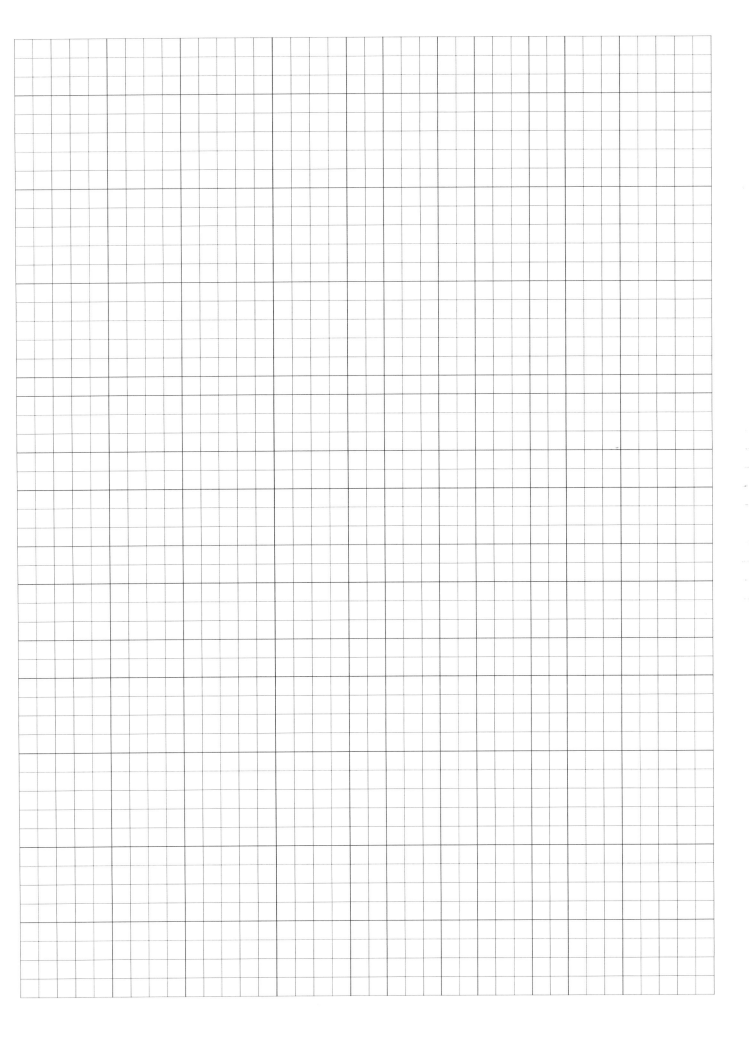

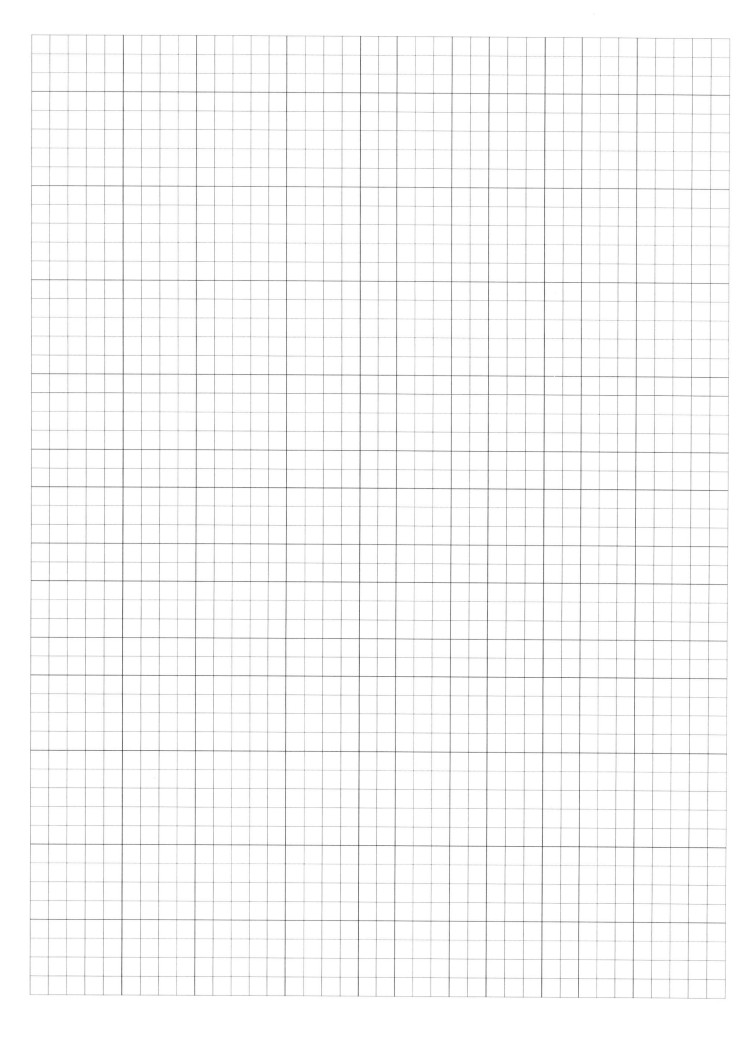

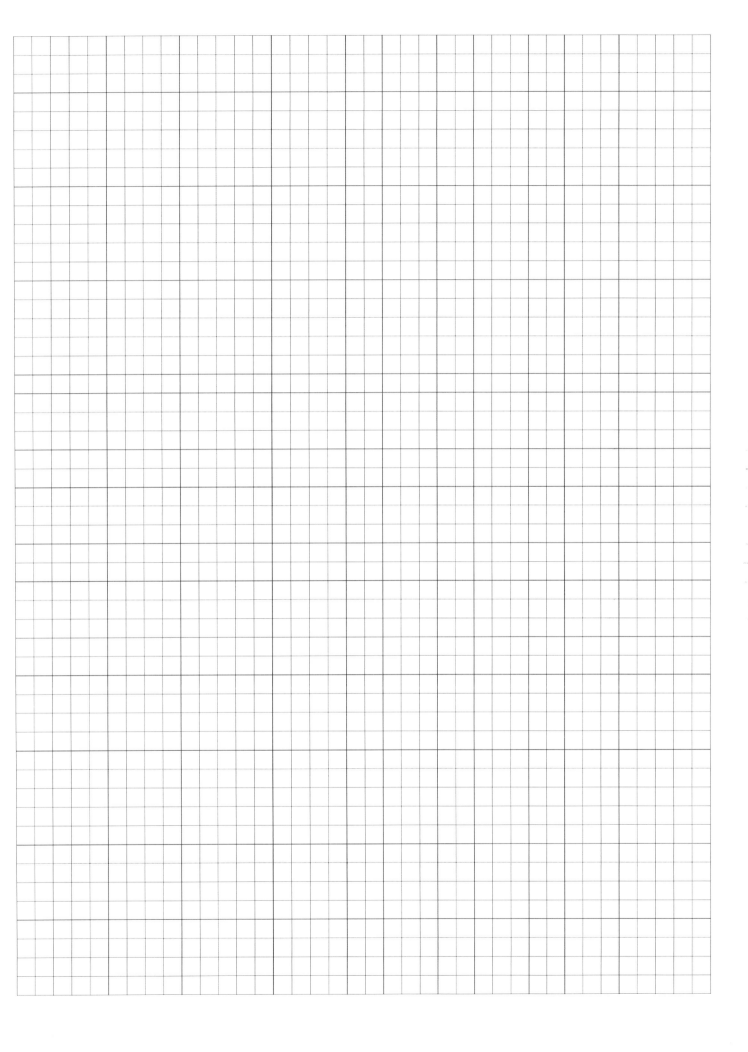

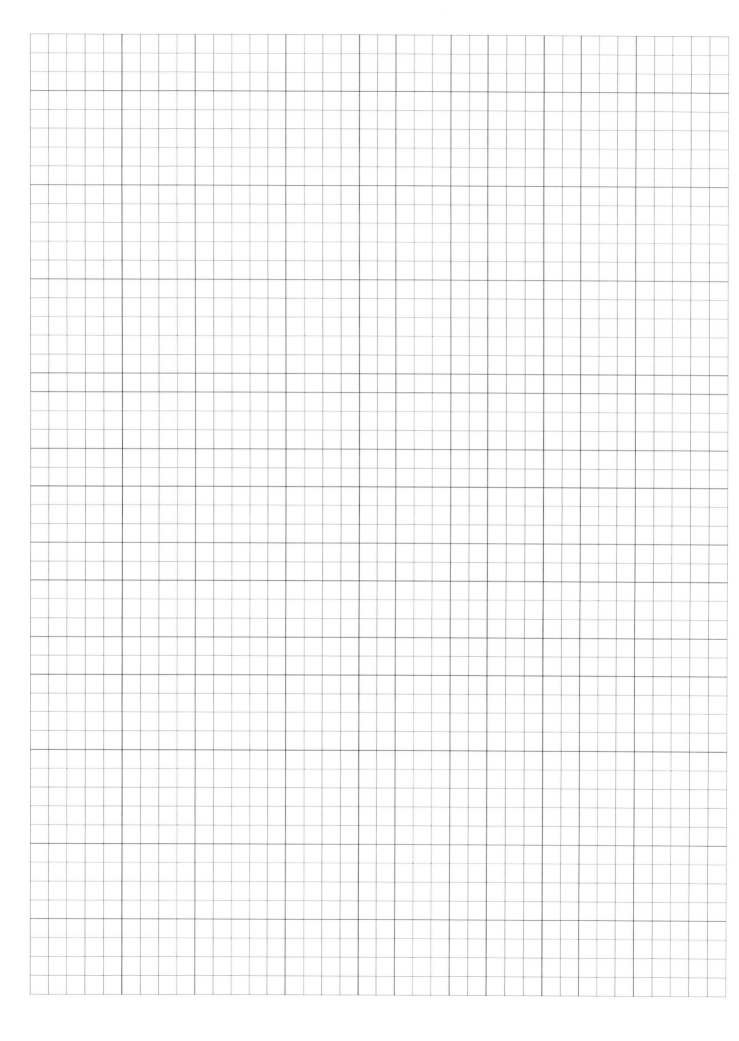

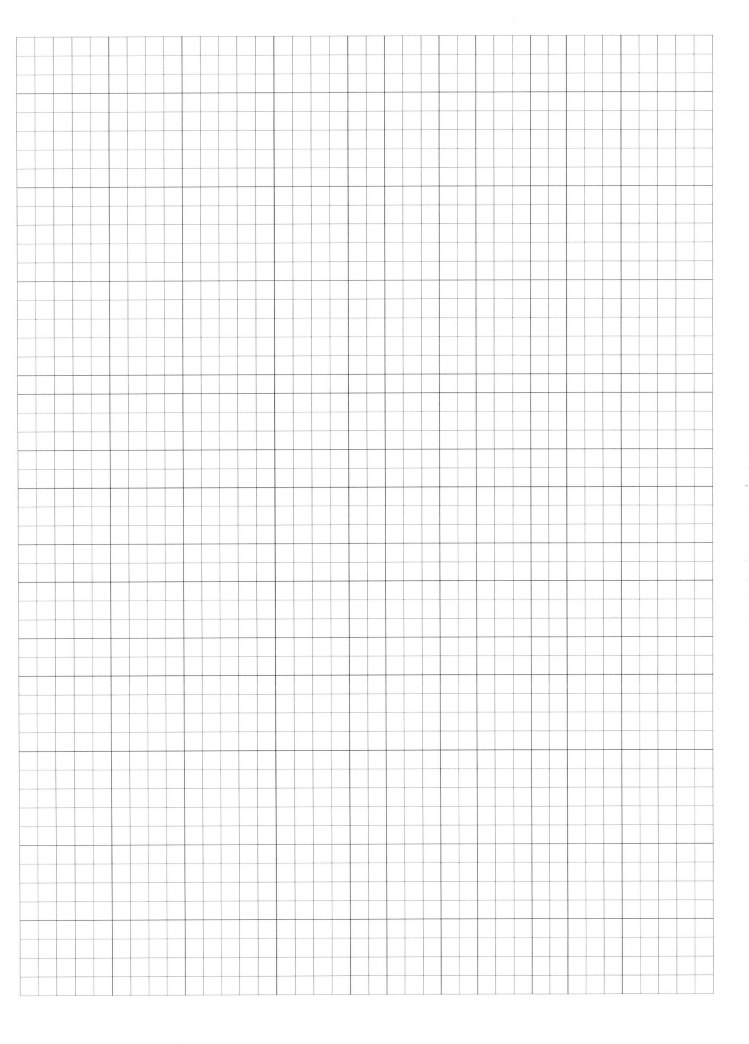

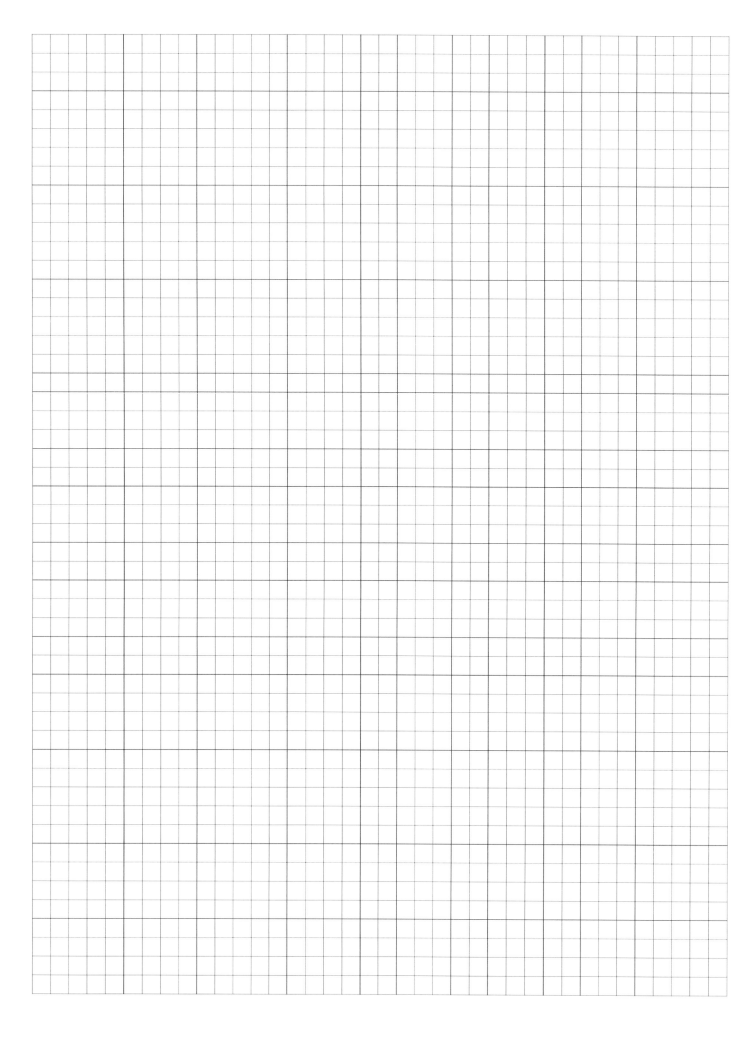

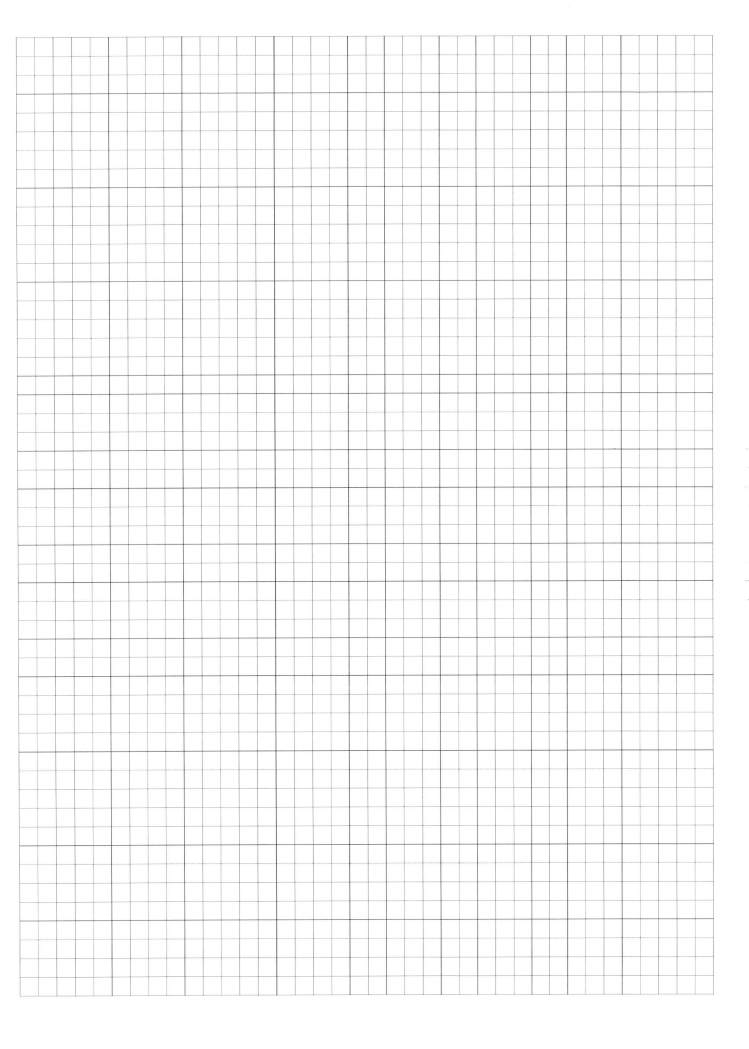

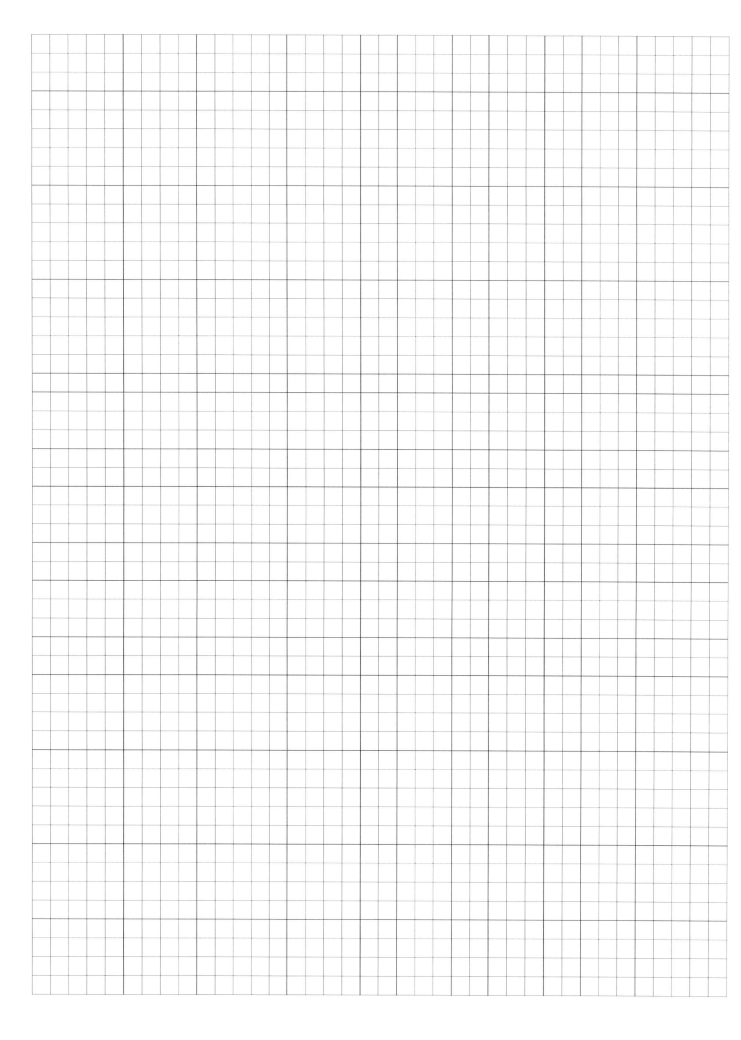

Made in the USA
Monee, IL
11 November 2021

81903489R00068